THE
READ ALOUD
FACTOR

THE
READ ALOUD
FACTOR

HOW TO
CREATE THE HABIT
THAT BOOSTS
YOUR BABY'S BRAIN

Rekha S. Rajan, EdD

CHICAGO

Published by Parenting Press
An imprint of Chicago Review Press Incorporated
814 North Franklin Street
Chicago, Illinois 60610
ISBN 978-1-64160-766-7

Library of Congress Control Number: 2022942666

Cover design: Jonathan Hahn
Typesetting: Nord Compo

Printed in the United States of America
5 4 3 2 1

This book is dedicated to the brilliance of Caroline Blakemore and Barbara Ramirez. Their visionary work in early literacy and as reading specialists continues to impact generations of readers, young and old

"All children should come into their magnificence
without limitations."

—Caroline Blakemore

CONTENTS

———

Part III: STEAM Reading: Bringing Arts and Technology into Your Read Aloud Habit 151

AUTHOR'S NOTE

WHEN MY AGENT approached me about writing a book about reading aloud, I was intrigued.

Reading aloud to a child is incredibly important, intimate, and life changing! I knew this because of my own experiences reading aloud to my three children, and the memories of snuggling in bed with my mom and younger brother as she read to us, every night (usually a Berenstain Bears book).

I didn't fully know the impact of reading aloud, because it was a natural part of my own childhood. Then I started doing my own research, exploring the different techniques and routines, and I read *Every Word Counts*, written by Caroline Blakemore and Barbara Weston Ramirez. The book they wrote is a phenomenal exploration of how and why we should read aloud, making this practice accessible to all parents, grandparents, and older and younger siblings. It was the catalyst and inspiration for this project, and I am grateful to them for allowing me to continue their original discussion on reading aloud.

Though many sections of this book are built from my own research and new, innovative studies of how reading aloud affects children's development (and our brain development through old age), I have also, at Caroline and Barbara's request, included and expanded on the framework they established for thinking about the stages of reading aloud from birth to 2 years old. Caroline and Barbara were pioneers in the field of early literacy—understanding, researching, and teaching that reading aloud begins at birth.

This framework of the stages of reading aloud has been greatly updated to reflect recent research and strategies that are important for a new generation of parents. Much has changed since *Every Word Counts* was originally published, including the types of books we read, where we read, whom we read with, and the ways in which we engage with technology while reading aloud. I hope that this book honors their original work and shows that the power of reading aloud will never diminish.

I would also like to express my deep appreciation and gratitude to Dr. Denis Evans, Dr. Charlie DeCarli, and Dr. Kumar Bharat Rajan (my husband!), who invited me to be a part of their research into how being read to at a young age supports better brain health in old age. This research is critical for furthering our understanding of why reading aloud to young children is so important.

In the past decade, much new research has emerged on how to read aloud to children, why we need to, and the ways in which it supports children's brain development, language development, and social-emotional development, along with various other factors.

I would also like to thank my editor Michelle Williams at Chicago Review Press for seeing the potential in the original proposal, managing editor Devon Freeny for excellent edits, and my agent, Lilly Ghahremani, for trusting me with this project. I'm also grateful to Lilly for coming up with such a great title! And I want to thank Mrs. Linda Lowery and Mrs. Margie Aker for sharing wonderful conversations on how reading aloud to children at home and in school has shaped their own lives.

As an arts educator, I also wanted to show how music and art are integral to the read aloud experience—a perspective that is new but vital to supporting a child's whole development. And because I have been raising three young children in a multilingual home, I felt it vital to recognize that our households have changed. We are not all from single-language homes anymore. Our classrooms and communities are filled with the languages of the world, and so our books and read aloud experiences should be too.

Perhaps the greatest change is how we once believed technology was detrimental to our children's development—raising my three young children in a pandemic showed me that in some ways technology actually gives our children an advantage. The quotes and anecdotes throughout this book are all real conversations or stories. One of my favorites is a conversation I had with a second-grade teacher who taught for nearly 40 years and always spoke of how you are "never too old to read aloud," a philosophy I share deeply.

I am so grateful to share this project and the exciting new research on how reading aloud changes the shape of our brains, as well as new ways for a new generation of parents (and parenting) to read aloud at home.

Thank you for reading.

INTRODUCTION

――――――

EVERY PARENT—NEW, EXPECTING, or seasoned—remembers that flood of emotions that envelops you, dizzies you, carries you, and cheers you on when you realize you are going to have a baby.

Now what?

Before you know it, you have a newborn who needs constant attention, feedings, diaper changes, hugs, kisses, burpings, baths, more diaper changes, spit-up cleanups, and so on. You wake up in the middle of the night for more of the same.

Or perhaps you spend most of your days chasing your toddler, keeping her from leaping off the couch, or playing hide-and-seek. Or your child is turning three, getting ready for the big *p* . . . preschool. Still, you have loads of laundry and are years behind on sleep, so it's common to focus on mere survival. And who can blame you? Research has shown that in the first three years of their child's life, parents are focused less on enrichment—that is, reading, spelling, math, and writing—and more on survival. (Of course, this focus can change in response to certain crises, such as during

1

the pandemic, when 75 percent of parents of young children felt a stronger need to lead educational activities at home.)[1]

How to prepare your child for kindergarten, much less *adulthood*, is probably not at the top of your list when you carry that baby into the house, bundled up in a car seat. When you can barely think about next week, five years can seem like an eternity! Before you know it, the months will go by; your precious little one will be walking, talking, going to preschool, and suddenly starting the first day of kindergarten. Just as your baby needs to be fed, held, loved, and nurtured, remember that you also play the most important role in shaping your child's language growth and brain development.

 Brainy Baby: You are your child's first teacher! The more you read aloud, the more your baby hears your voice. And the more your baby watches you read, the more your child will love reading too!

Research has shown that children are watching adults and their reading habits,[2] and that the more books in the home, the more likely a child is to read and want to participate in a read aloud experience.[3] I love to compare this to how my husband drinks carrot juice every day. Before this was a trendy drink, our oldest son (Jagan, who was four at the time) would watch him and then ask to drink from Dad's cup. He was the only child in his preschool who would bring carrot juice for shared snack day. Many children (and their teachers) asked what this strange drink was, tried it, and ultimately threw their cups away. What we know is that our children are watching us and will model our behaviors, whether it is cleaning, reading, or drinking carrot juice.

What is important to know and focus on is that from birth (and even prenatal development) through age 5, children's brains are developing at an exponential rate. Every activity you do at home—such as the various ways that you talk, sing, or play with your child—is helping his brain grow. And one of the most important activities is reading aloud.

When you read aloud to your baby in utero (before he is even born), his brain activity increases.[4] The experts at the childhood development nonprofit Zero to Three state that "a child's brain undergoes an amazing period of development from birth to age 3—producing more than a million neural connections each second."[5] The Centers for Disease Control and Prevention (CDC) expand on this statement, noting that the "first 8 years can build a foundation for learning, health and life success."[6] What we do as parents and the choices we make influence our children because they are always watching us, hearing us, and wanting to be like us.

Being a Parent Is Being a Teacher

In these early years, even though you may not realize it, you will be the first and most important teacher your child will ever have. This is the *most* important period of your child's brain development.

Yes, your child will go to school, interact with peers, and learn new and exciting subjects. But no teacher has the power that parents have to ensure academic success. And no teacher has the early access you do to stimulate your child's brain growth. What is this power? It's the power to give your baby the gift of words. That's right. Words! Short words, long words, common words, and uncommon words.

Lots and lots of words, every day.

Brainy Baby: Building language starts with words. Talk to your baby every day about everything you are doing. Your baby hears all the casual conversations in your household: "I'm making broccoli for dinner." "What time do you have to go to work tomorrow?" "It is so hot out today!" Your baby is always listening to you.

Recent research tells us that what determines future academic success is the amount of words per hour babies hear before the age of two.[7] It is these early life experiences that are *critical* for supporting children's development by preparing them for adulthood.[8] That means when you read aloud every day, you are helping to build your child's language and vocabulary.

Exposing children to new adventures, hands-on activities, and social interactions is a big part of raising a happy, healthy child, and many of these adventures begin even before your baby is born. So you may be wondering: If this is so obvious, why isn't everyone reading aloud from day one? Why continue to write more books and update editions of texts that tell us to read aloud?

Well, not every parent knows *how* to read aloud even if we all know the *why*.

Is it OK if your baby chews on the board book? *Yes!*

Is it OK if your child crawls away while you are reading aloud? *Yes!*

Is it OK or helpful to read aloud when your newborn is asleep? *Yes!*

There are many ways to read aloud to your child, along with different texts, new technologies, and books that are published in dual languages to support the growing diversity in our communities and schools. Reading aloud is important, but the ways in which we read aloud and are comfortable reading aloud have changed.

I came from a family of readers. When I visited my grandfather in India, he always had a book in his lap, laid across a pillow. I assumed this was normal behavior. I didn't realize that not everyone read aloud, read books, or visited the library for fun. Until I became an adult, I never realized how many people are afraid of reading aloud or mispronouncing words, and are uncomfortable talking about it.

For example, one of my friends (who was in her early 20s at the time) had her first baby, a boy, a few years ago. As a gift, I gave her a book about dinosaurs to read to her son. It was my boys' favorite dinosaur book, with large, bright pictures and little buttons on the side with pictures of dinosaurs and their corresponding sounds. I was so excited to share this book.

My friend loved the book as well, exploring it and pushing buttons, the sounds of a triceratops and *Tyrannosaurus rex* surrounding us. Then she looked through the book and said to me with a laugh, "I don't know if I can read this book to him. I probably won't be able to pronounce most of the dinosaur names!"

That moment was profound for me as an educator and parent.

It made me ask, How can you read to your child if you are not comfortable with the content? How can we encourage new generations of parents to read aloud? How can we find ways to support reading aloud when libraries are closed (during the pandemic)? How is reading aloud different when everyone is on a smart device?

And does it really matter if you pronounce every dinosaur name correctly? Wouldn't a toddler be excited to just roar with every page? Hearing my friend's concern made me realize that some parents worry that by reading aloud the "wrong way," they are teaching their child the wrong words, ideas, or pronunciations. And then I wondered, Is there really a "correct" way to read aloud?

Reading and writing skills begin at birth, when your baby is first exposed to language. Learning to read doesn't start when a child goes to school and gets her first phonics lesson. It only comes easily when children have been immersed since birth in the world of words through a steady process of hearing family read aloud. That means that there isn't any right or wrong way to read aloud.

It's like I told my friend: It isn't about pronouncing every dinosaur name correctly. By reading aloud, you are teaching your baby that something *is* a dinosaur, or rock, or tree. The images and words in the book help your baby to learn language skills.

 **Reading Rocks:** Not all of us are paleontologists, but we can teach our baby what a dinosaur is just by saying the word out loud, pointing to the picture, and describing what it looks like.

When you think about it, most communication and everything you learn in school involves words. Words are the basis of literacy—the ability to read and write. To succeed in school, children need to pay attention, listen, focus, understand, and communicate through writing and words. These basic skills form the building blocks of literacy. Your baby acquires these building blocks naturally in the first years of life, but only if you set aside time every day to lovingly read and talk to your baby.

As a caregiver, you need to begin early, talking and reading to your baby during prenatal development and pregnancy. Your unborn baby hears you—your voice. And it's not just your voice, it is anyone in the household—siblings, extended family, grandparents—who will help to reinforce language development and literacy every day. In fact, researchers have noted that after the mother's voice (which is the loudest in the womb), outside male voices are actually heard in a louder tone in utero than outside female voices. Enter dads, uncles, grandpas, and so on.[9]

After his birth, you should continue talking and reading to your baby—even if you think he is asleep and isn't listening. By three years of age, your baby has been absorbing the language in his or her environment at a pace and intensity that only happens in the first years of life. You might think your baby is exhausted, weary, and just trying to adjust to this new life after birth, but his brain is working overtime, preparing for an incredible life journey.

 Brainy Baby: Your baby's brain grows the fastest it ever will in the first three years of life. Reading and talking to your baby every day is the best (and only) way of supporting language development during this short period of time.

I wrote this book with the hope that we all will read aloud to our children. Reading aloud should not be a daunting, overwhelming experience but one that fosters bonding, care, and empathy, and encourages children's brain, language, and social-emotional development. I argue that reading aloud *should* be:

- A top priority for parents
- A natural part of how we talk about parenting and family routine
- A simple habit for parents, especially tired new parents (trust me, I have been there!)
- A first-choice activity, even for parents who may not be big readers themselves

Many people know they *should* read aloud, but they don't have a game plan or strategies for how to do so effectively.

Remember that it is you, above all, who are in this child's life, and the behaviors you model are an important part of demonstrating why reading is important. If you always have a book with you or near you, in your purse or bag or in the car, or if you read for pleasure in the evening, your child is watching and emulating those same behaviors.

What Does the Research Say?

Decades of research show how reading aloud supports children's development. Reading aloud to a child enhances and promotes language acquisition and is connected to later academic success.[10]

The following data outlines how by reading aloud you expand your baby's knowledge of words. This data shows how many words your baby would have heard by the age of five if she were read to every day or not at all:[11]

> Never read to: 4,662 words
> Read to one to two times per week: 63,570 words
> Read to three to five times per week: 169,520 words
> Read to daily: 296,660 words
> Read to daily with an average of five books a day: 1,483,300 words

You are probably wondering, *How is this possible?* It doesn't seem realistic that a child who is read to daily can have almost 300,000 more words in his vocabulary than a child who is never read to. Researchers call this the "million-word gap": the simple experience of reading aloud to your child expands her vocabulary by a million words by the age of five. Many factors are at play here.

Studies have also found that over 70 percent of parents are reading aloud to their babies before their first birthday.[12] What is fascinating is that the percentage of parents who read books aloud at home has continued to increase over time. Why is that? There could be many contributing factors (quarantine perhaps being one in recent years), but the American Academy of Pediatrics released new guidelines in 2014 encouraging families to make reading aloud a routine at home. Think about this for a moment: the same organization that develops guidelines for what foods your child should eat and what physical activities will help your child stay healthy is now including *reading aloud* as part of a necessary component of your child's development.

For many, this recommendation is easier said than done. Having access to books, recognizing that multiple languages may be spoken at home, and finding ways to support parents who were never read to as children and therefore don't know how to engage with their child (that's my dad, for example) should be addressed in these recommendations.

In my own childhood, my mother read to us every day because her father modeled the importance of reading and reading aloud. My father, on the other hand, was never read to as a child and so never read to us. Reading wasn't an important part of his childhood.

Reflect on your own childhood or even those of your parents. Was reading and reading aloud an important part of your routine? Many factors influence our understanding of reading and books, and much of that begins with access. These facts aren't meant to be a criticism of parents who don't read aloud, have never read aloud, or have never been read to. You're here after all, reading about why this is important and what to do about it. We are all working and grinding out our careers, family life, and mental and emotional health, and coming to this resource now is a step you're taking to work on this important piece.

Even as I write this book, I must acknowledge that I read aloud more with my first child and less with my youngest. I did, however, encourage my children to read aloud to each other, and that experience is just as valuable as being read to by a parent (if not more valuable). This scenario is true for most families. Though around 30 percent of parents say they read aloud to their child from birth, this decreases over time, including how long parents read aloud (usually less than 15 minutes).[13]

I see a great opportunity here to recognize that these early life experiences are critical for supporting children's cognitive and emotional development. The home environment provides the child's first literacy experiences. Reading aloud to a child enhances and promotes the child's language acquisition and is connected to later academic success in so many ways. Specifically, reading to children at an early age is connected to language growth and reading achievement.[14]

The American Academy of Pediatrics also breaks down the recommendations for reading aloud into stages for emergent literacy (that is, the development of literacy skills such as reading) beginning after birth and continuing through kindergarten. In fact, researchers are now arguing for reading aloud through middle school and recognizing that reading aloud isn't just the parent's job. Older siblings (such as those in high school) can and should read aloud to younger siblings in the home.

 Brainy Baby: Building vocabulary starts from birth. The start of reading aloud is knowing how you talk to your baby every day. Do you sing to your baby? Imitate his cooing sounds? Or just say I love you? All of that counts and is important for brain development.

Children who are read to, and who read aloud to others, gain a broader vocabulary[15] and develop strong oral language skills.[16] The first five years are considered the most critical years of brain development. In fact, the developmental stages between preschool and second grade are the most important times for preventing and reducing reading difficulties in children.[17] Scholastic's *Kids & Family Reading Report* found that the more often a child is read to, the greater the child's reading fluency.[18] This is true for not only children who speak English but also children from households where multiple languages are spoken.

Reading also supports your child's bilingual language development. In a study exploring how to increase vocabulary in children who speak both Spanish and English at home, researchers found that reading aloud supported children's language development in both languages. Perhaps most important, researchers concluded that "parents can learn how to enrich their communication with their children using read alouds with minimal training."[19]

 **Reading Rocks:** Reading in your native language or books that are bilingual will support your baby's language development in equal parts. It doesn't matter what language you are reading in as long as you are reading to your baby.

Reading aloud is also an important way to bond with your baby, build strong social-emotional skills, and develop empathy for others. Yes, there are multiple benefits of reading aloud. But the most basic is that by reading to your child every day, the shape of the brain actually changes to become bigger, stronger, and healthier throughout life.

Early life experiences, specifically reports of more frequent participation in cognitive activities, have been shown to be associated with higher cognitive function[20] and slower age-related cognitive decline.[21] Researchers continue to explore how reading aloud actually stimulates and actuates your baby's brain more than other activities and how reading aloud changes the shape of the brain.[22] However, it is not clear how these early life experiences are associated with biological measures of brain health in old age, or how these biological measures relate to cognitive performance.

This means that although we know that what we do as young children makes our brain stronger and healthier in old age, we still don't know which specific activities are most effective in supporting brain health. But we do know that reading aloud supports a healthy brain.

Why we should be reading aloud to our baby will always come back to how it affects the brain. In my ongoing research on brain development and reading aloud, my colleagues and I found that among our research participants who were all older adults (over age 60), those who said they were read to every day as a child had a left frontal cortical volume that was 1.82 cubic centimeters greater in old age than those who reported being read to once a year or less.

Our conclusion? When you read aloud to your baby every day, your baby's brain gets bigger! The connection to reading aloud and brain development is unmistakable. And the connection grows over time. We found that early life experiences, such as reading aloud to your child, are associated with improvements in left hemisphere brain volume and increased cortical thickness. This suggests a healthier brain through childhood, adulthood, and old age, resulting in better cognitive function in old age.

Obviously, many genetic and environmental factors come into play, but consider how often we have heard about older adults or even our elderly family members who struggle with memory loss, dementia, and brain decline in old age. Could it be true that being read to frequently as a child can actually change the shape of your brain?

Absolutely.

This brings me to the exciting, innovative research that motivated this project with the purpose of sharing new data on how reading aloud supports us beyond the early years. My colleagues and I used data from a large population health study that took place over two decades (and is still ongoing).[23] The study began in 1993, when participants engaged in multiple cognitive tests and baseline interviews, among other forms of data collection.

During the baseline interviews, participants reported on three early-life cognitive experiences: being read to, being told stories, and playing games. Participants responded to the question "How often did someone in your home read to you as a child?" with five possible responses: "every day or almost every day," "several times a week," "several times a month," "several times a year," or "once a year or less."

These five response options were also used for questions about two other early-life cognitive experiences: "How often did someone in your home tell you stories when you were a child?" and "How often did someone in your home play games with you when you were a child?" Responses were assigned values from 1 (low) to 5 (high) and used to examine their association with brain indices in old age.

The frequency of being read to as a child was positively associated with 5 of the 86 volumes, all on the left hemisphere of the brain, and 16 of the 62 cortical thicknesses throughout the left and right hemispheres of the brain. Being told stories and playing games also supported brain size and health, though not to the level of the strong, positive benefits of reading aloud.

The MRI figures of an adult brain on page 14 show how being read to as a child actually *changes* the shape the brain will have in older age. The shaded areas indicate how and where the brain shows increased volume and cortical thickness. Our results show that being read aloud to at a young age is associated with increased brain volume and cortical structure.

Imagine your brain is a sphere: *brain volume* corresponds to the *inside* of a sphere, and *cortical thickness* to the *outside thickness* of the sphere.

Early-life reading experiences (being read aloud to) significantly improved the outside thickness of the brain in as many

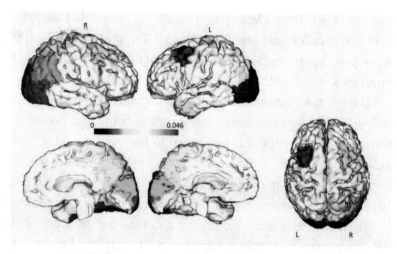

Brain Cortical Thickness

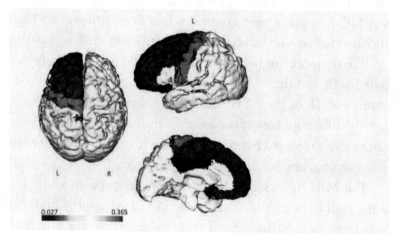

Brain Volume

as 16 regions. We found that a higher frequency of being read to as a child was associated with left frontal lobe volume. We also found that it is important not only to read aloud but also to read aloud *often*. Reading aloud to your baby for just 20 minutes every day will help the brain grow bigger and healthier for your baby's entire life. That means the *more often* you read aloud, the better.

Think about that for a minute. What you do now impacts your baby for over 60 years. Early life experiences have the potential for a long-lasting impact on brain health through increases in volume and thickness of the cortex that last into old age. More important, the home environment and home-based interactions supported sustained development and better brain health in old age.

Our findings suggest that reading aloud to children at home is a valuable and important part of supporting sustained brain development across the life span and better brain health in old age.

It's important to note here that reading aloud isn't "curing" any neurological or brain-related matters in old age. In the same way we are encouraged to engage in "brain-strengthening" activities as we age (such as crossword puzzles or sudoku games), the ways in which we were read to as a child directly influence our brain health. This isn't a quick fix or get-a-stronger-brain-as-you-age trick. This is something that is part of lifelong healthy development for your child.

The idea that reading aloud supports children's development isn't new. We know that you should read aloud, encourage your child to read, have books, and support literacy skills. The excitement and motivation around the findings in this research project show that there are immediate and *lifelong* benefits to reading aloud. The time you take to read aloud to your baby every day affects brain development, language development, and social-emotional development as your child grows, setting your child up for a life of success.

In our current times, it is even more important for us to have the tools, techniques, and knowledge needed to read to babies. There are emotional benefits to reading aloud and being read to as well. Researchers found that children in a pediatric intensive care unit demonstrated reduced levels of pain and stress when they listened to a story out loud.[24] In this study children ages 4 to 11 were split into two groups: one in which they were led through a series of riddles and games and the other in which a storyteller read aloud a book.

While both groups demonstrated a decrease in stress and higher levels of oxytocin (a "feel good" hormone), the children who were read aloud to showed a decrease in pain that was almost twice as much as the children who played games.

Here's the thing. It is not just you who may be struggling. Parents all over the world have been grappling with the challenges of virtual learning and supplemental activities at home. How and when we read aloud and engage with our children is constantly changing. Parents of multiples might find that survival each day is a greater focus than finding time to read.

Here's an example that I want to share because even though I love being a mother of three amazing children, I have never had the experience of balancing twins or triplets. But I have seen other parents do so. When my sister-in-law had twins, I honestly thought that her read alouds with her first child were going to completely disappear. Instead what changed was that my niece was invited to sit between her sisters and read aloud to them every day, even if it was for a short time. And while these three little girls were sitting together (two of them possibly sleeping), it was a connected read aloud experience.

If you find one baby is napping while the other is alert, that is a wonderful opportunity to read aloud and build a special bond with each of your babies. Remember that even a few minutes of reading aloud is better than none.

One aspect that is always consistent is the role you play in your baby's development. What I argue here is that reading aloud is not just something that affects your child's readiness for school. It's not only a benefit to him in old age. It's also a cozy bonding activity, a shared time of connecting, laughing, and experiencing new worlds and characters together.

Research has even found that children really do love being read to. In a study on children's attitudes toward being read to, over 70 percent reported that read alouds made them feel "happy," "relaxed," and "good inside."[25]

If there is even one activity that you can do for your baby to help strengthen her brain, and it is one as simple as reading aloud, why not? It is the opportunity to create a shared experience, teach empathy, and build language development. It is a part of your child's health and wellness routine that he will benefit from and enjoy for the rest of his life. You will increase her language development and brain size, and you certainly will increase the size of her world through reading.

It's a fact.

The Purpose of This Book

I have spent decades working in schools, teaching preschool through elementary grades, and what I have recognized is that year after year, the kindergartners and first-graders who enter classes are:

Enthusiastic

Happy

Excited

Scared

Bright-eyed

Nervous

Curious

Creative

But very, very often they are illiterate.

Many children come to the classroom with little or no experience with books, and as an educator I have found that it is because families either don't have access to books or believe that it is the school and teacher's responsibility to teach the child to read. In many cases children entering my classroom without being able to recognize the alphabet left my class at the end of the year as avid readers. That was my role and responsibility as a teacher.

But is that really fair to the child?

The National Center for Education Statistics notes that only 2 percent of children entering kindergarten can read by sight (more on sight words in chapter 2). That means 1 in every 50 children enters kindergarten with reading skills.[26] In most cases the children never had someone in their household read aloud to them. The reasons are many. Some of these reasons are even more real and important right now.

When virtual schooling became a reality, parents around the world grappled for new ways of engaging their children, ensuring that learning didn't decline (even if it did), and finding resources to support homeschooling. Through all of this, reading aloud has completely changed. How can we expect parents to read aloud if we don't consider access, ability, and equity?

- Access to books: How can you read aloud if you don't have books in your home? Are libraries still functional? What other resources are there to support you as a new parent?
- Ability to read: How can you be confident in your own ability to read if your first language isn't English? How can you support language development and the development of multiple languages through children's books?
- Equity: What characters do you want your baby to hear and learn about in his first books? Those that look like him? That help you show your baby your culture, family life, and values?

Those factors influence *how* we can read aloud and *what* we read aloud. Even if we know the why, not everyone has a household filled with books. I didn't have a household of books growing up. When we think about how to access books, we often think of the library.

While many organizations shut down during the pandemic, libraries found a way to expand access to books and resources, becoming a lifeline for families during a very difficult financial and emotional time.[27] Many librarians also lost their jobs and sought other ways to support the community by providing outreach and literacy support. Facing the need from the public, libraries began to offer more e-books, added curbside pickup, and even curated free art supplies and packages to help struggling families.

The issue of access to books is not just for parents of young children or those in communities with a lower socioeconomic status. In fact, the pandemic brought forth the challenges within schools and higher education. Many students at two-year community colleges didn't have access to books or the Internet.[28] How were they supposed to be successful in their academics?

This impact trickled down into schools in states that reallocated funds and removed school librarians from their roles completely. When we were in California prior to the pandemic, it was dismaying to see that my children's public elementary schools did not have a school librarian at all. Though the classrooms had books, the experience of entering a library and taking ownership of your book choices was something my children (and many others around the country) did not have the opportunity to experience.

The library is more than just a place to borrow books. It is a resource center rich with opportunities for expanding learning both within school and in our homes. We found this more than ever during the pandemic's widespread lockdowns. People who had never utilized the online services that many public libraries offer, such as OverDrive or Libby, were suddenly downloading and checking out books at an incredible rate. Public libraries in California and across the country noted a 50 percent increase in digital downloads during the pandemic.[29]

But these resources existed before COVID. Libraries have also long served as a safe space for the public and for children who may need a refuge from troubled home environments, as well as a quiet area to study and interact with peers.[30] Many families have, and continue to, utilize the free Internet offered at public libraries. I have talked with many parents of young children about how they miss the free and weekly programming offered through the public library—*accessible* and local opportunities for enrichment and socializing that quickly disappeared.

We have to consider this: What do we do if the world flips upside down again? How do we continue to find ways to support our children's brain development when we don't have access to books, or the ability or even time to read aloud? How do we build a literacy-rich environment with books that reflect our own home and the way we look?

As a teacher and a parent, I did not want this book to be stifling, hard to read, or filled with research that was difficult to dissect and understand. Instead I hope this book offers a new approach that gives you the most updated information on the importance of read alouds, guidance on how to read aloud, and a clear explanation of how reading aloud prepares your child for school and can foster a love of lifelong learning and reading. Many sections of this book will talk about reading aloud to your baby and building upon the stages of reading aloud that are critical for developing and establishing a read aloud routine, while other sections will include anecdotes, ideas, and strategies.

Though I focus on reading aloud to infants and toddlers in some sections (because the earlier you start, the better), I also include the ways to read aloud (and support read aloud experiences) with older children. I focus on developing and building that habit early on because though it may seem like your baby isn't listening, she is. Though it may feel like we are too busy (we are), your baby treasures that special moment of reading aloud and bonding.

I hope that you will find that reading aloud not only gives your child a sense of well-being but also provides the underlying neurological nourishment for language development and future academic success. It is my hope that through this book, you will have the information to support *why* you should read aloud, ideas and strategies for *how* to read aloud, and the foundation for choosing *what* to read aloud and identifying *where* and *when* to read aloud by developing a treasured, daily read aloud routine. And remember, through all of this, you are helping to give your child, or the children in your life, the best possible start in education and life.

How This Book Is Organized

This book is divided into three parts that take you through understanding *why* you should read aloud to your child, *how* to read aloud to your child, *what* to read aloud to your child, tips and techniques for gathering resources, stories of successful read aloud experiences with renowned educators, and suggestions for *where* and *when* to read aloud with your child or as a family.

Part I explains *why* it's important to nourish your child's brain with words, language, and interactions. I'll share what I have learned over decades of experience and the most important benefits of reading aloud to your child, showing how literacy begins at birth. I will do my best to explain how the effects of reading aloud influence future reading and learning ability and the important role of parents, caregivers, and siblings in reading aloud. Finally, I hope to show some ideas and basics of reading aloud and how to prepare for your daily read aloud habit, whether it is after school, in bed at night, during breakfast, outside, or even during bath time.

Part II presents ideas for *how* to make the most of read alouds starting at birth. These chapters expand on read aloud stages in early life, provide STEAM (science, technology, engineering, arts, and math) activities that you can do at home, list ways to create a read aloud routine, and share tips and techniques on how to make reading aloud most effective. Part II also builds on why reading aloud is (and continues to be) so incredibly important for your child's development. Each chapter also includes sections on why reading aloud will change as your baby gets older and suggestions of various books that will help your baby's brain, language, and social-emotional development as she grows.

Part III further explores what types of books are best for read alouds. Here I integrate how STEAM topics are naturally a part of children's books and can be utilized to support your reading aloud. STEAM (or sometimes only STEM) topics are very much a part of grade school curriculum. No longer are subjects taught independent of one another; rather, learning is a joyful blend of all different subjects. This is to show children how the world is connected and to bring value to all the subjects we learn in school; we emphasize this when we use art and music to build a quality read aloud experience.

I also expand on how to use technology (tablets, e-readers, smartphones, television) effectively. This was one of my favorite sections to write. At some point we have to acknowledge that those of us with children spent an enormous amount of time on video calls and devices during the pandemic (even those of us without children). What is beautiful here is that I have seen grandparents doing read alouds with their grandchildren over FaceTime, singing songs together, sharing favorite moments, and writing stories. And what emerged through the challenges was a new type of lovely bonding experience. So rather than shun or disapprove of technology that we use every day, this section summarizes some of the latest and emerging technologies for supporting reading aloud (devices can now read to your child) and how best to use these devices.

In this section I also share resources for you, as a parent, caregiver, relative, or friend (or even older sibling), to build a diverse and rich home library of books to read aloud, and I address our shared concerns as parents of what, where, and how to read aloud in our changing, diverse, and technologically focused world. I also give a brief summary of how to raise a literacy-rich child, some final tips, and some words of encouragement.

Although in this book I use the phrases "your baby" and "your child," I certainly acknowledge that the person doing the read alouds could be a grandparent, a loving caretaker, or a sibling. As a mother of three young children, my instinct is also to make everything plural, so I certainly acknowledge that you may have one or multiple children climbing on top of you (or jumping or spitting or biting) as you read this. I am also conscious of recognizing gender identity and equality, and I chose to use the terms *baby*, *child*, *parent*, and *caregiver* as opposed to gender-specific terms.

This book is meant to be inclusive and equitable as I share my ideas and focus on the experiences, processes, and bonds that are created when reading aloud, through the private moments of hearing each other's voices, while recognizing that everyone's family, environment, and situation is different, unique, and valuable.

I have to state that though there are stages of reading aloud and physical and language development for your baby, the stages for reading aloud are all somewhat generalized based on recommendations and guidelines from the American Academy of Pediatrics. My three children's speech, language, and physical development all varied to the point that I sometimes wondered what I was doing right or wrong, if anything at all.

How could my first child not speak until he was four, my second child talk to me in full sentences at one year (and potty train herself), and my third child sing to me in multiple languages as I got him ready for bed each night? This is my Jagan, Madhavi, and Arjun, whom I will refer to throughout this book as examples of stories and anecdotes.

The lesson here is that each child, parent, and family dynamic is unique, so please take these ideas and activities and suggestions as just that—a framework for you to see what works best for your family and your baby (or babies).

I want to share that when this project came to me, it was originally to update and revise a book that was already in print. I was intrigued. But so much has changed in how we engage with our children, our understanding of how our children gain literacy skills, and how we read aloud together. Every decade brings new changes, resources, and technologies—and reminds us of what we experienced and what we want our own children to experience.

Or as my kids have asked, "Mom, did you have a library in 1880?" *Sigh.*

This book was written for us, the modern parents, who have smartphones and tablets, work full-time and try to manage a household, draw support from other family members and friends, and are doing our best. I am that parent too, and I wrote this book as a way to support families in developing a read aloud habit, recognizing that no one way is perfect, and finding that reading aloud is an accessible, meaningful way to support your baby's development.

I also shift focus between reading to your baby or child and supporting a lifelong love of reading and reading aloud. Yes, it is important to start reading aloud from birth and onward, but all of this is to prepare your child for school and for socializing and interacting with peers. It is also to encourage a child who may not speak English as a first language, to help children build imaginative thinking and empathy, and to set children up to thrive by creating a literacy-rich home environment.

My goal is that in the near future all children will come to school having been read to *daily* in the first five years of their lives (and beyond). I also hope that this book highlights the challenges we face in this country with reading and literacy, and that literacy skills will be recognized and supported in schools even more. And with that, we can collectively promote the understanding that all children deserve the opportunity to become proficient readers, build their language skills, and explore their own selves and identities through books.

Let's start reading.

PART I

WHY READ ALOUD?

HOW READING ALOUD SUPPORTS YOUR CHILD'S DEVELOPMENT

AS PARENTS WE observe how our babies listen to us, how they tilt their heads toward our voice, the glimmer in their eyes as they recognize our voice, their first laugh. Parents start talking to their baby before birth, beginning the interactive dialogue that will later turn into give-and-take conversations.

Did you know that within four months, babies not only know the sounds of their parents' spoken language but also recognize their own names?[1]

While we know that reading aloud to children stimulates brain development, experts note that your baby's speech and language development is "also enhanced, and the experience of reading aloud enriches the family experience and contributes to social/emotional development."[2]

There is no greater and more direct way to support a child's language development than by reading aloud. Based on decades of research and my own studies of how reading aloud supports brain development, I developed a framework of three keys areas of development: overall brain growth, language acquisition, and social-emotional health.

These are all important parts of your children's growth, and they are all interconnected. Your child's brain gets bigger as he learns new vocabulary (language) and understands how to communicate with his peers (social-emotional growth).

Yes, supporting your baby's physical brain health is incredibly important. (It's part of why we take those prenatal vitamins.) However, in the past few years, early childhood educators have increasingly focused on understanding and supporting children's social-emotional health and well-being. Much of this can be attributed to the pandemic and the uncertainties that came with wearing masks, social distancing, and trying to explain why we were isolating ourselves.

It can also be promising to acknowledge that social-emotional development is a key area of learning and growth for young children. For example, your infant communicates with you through social cues and emotional connections: cries and laughter, winks and nods. The exciting part is that you can help to foster growth in all of these areas of development through reading aloud books.

1

BENEFITS OF READING ALOUD FOR OVERALL BRAIN GROWTH

THE BRAIN IS such an interesting, vital, amazing organ. How does it control our entire body, helping our muscles to function, creating connections among brain cells that engage us in thinking, analyzing, and exploring the world?

Researchers continue to learn new things about how the brain develops and supports our own functioning and learning. Consider that just a couple decades ago, there was very little information on prenatal brain development. Could your child hear you while you were speaking to your growing belly? Yes! This is how your child recognizes your voice immediately after birth. Is the brain fully developed at birth? No! The brain continues to grow, expand, and change shape after your baby is born.

In fact, researchers identified that between the ages of 0 (birth) and 3 is the time of greatest brain growth and development. This very short time period affects not just the size of your child's brain (and its functioning) but also how quickly your child can learn new skills (both independent and social), acquire multiple languages (receptive language), and develop literacy. Early life experiences are crucial to supporting your baby's brain development.[1]

In a 2015 study of the relationship between reading aloud and brain development, researchers found that when parents read aloud to their preschoolers, their child's brain became activated in the areas that support visual imagery and language development.[2] Researchers have also found that when children are actively engaged in read alouds, their brains become "turbocharged"— neural connectors in the brain are expanded, amplified, and working overtime to help your child develop skills in comprehension and language development.[3]

Babies come into the world with about 100 billion brain cells.[4] While the brain is constantly developing and changing and growing, the activity in the brain is affected by the number of signals and connections that are taking place.[5] Every time you speak to your baby, sing with your baby, and read to your baby, she is listening and increasing brain development. You may not even realize how often you speak to your baby. Isn't it amazing how directly our natural instincts in communicating with our babies affect their development?

Researchers generally use the terms *brain development* and *cognitive development* interchangeably. What is cognitive development anyway? Cognitive development is how your baby processes information and gains knowledge. It is also what is studied among older adults who have cognitive decline, which can lead to dementia or Alzheimer's disease. Brain development is how quickly your baby's brain grows and develops. I use these terms interchangeably in discussing how your baby's brain will grow bigger and healthier through activities that support cognitive development.

Promoting Listening

Anya looks down into her baby's eyes as she cradles Krishna in her arms. "My sweet little boy," she sings as she rocks him, and he looks at her, listening to her words, her familiar voice. "Pyaara beta," she sings again, this time in Hindi.

Does baby Krishna understand exactly what Anya is saying? Yes! He hears the inflections in her voice as it moves up and down. He sees the soft expression in her eyes. He feels her rocking him, and he is listening. He is *listening*—not just with his ears but with his entire body and senses. He knows that his mother is speaking to him, and he knows that she loves him through all of these actions, sounds, and words.

Every language is different and unique—in the inflections, the speech patterns, the tones used for certain words and expressions. Yet babies all over the world quickly become experts in their own language. For example, in French the words *le maison blanc* literally mean "the house white." In English you would say "the white house." How do babies learn the correct word order of their own languages? They learn by listening to their families.

Listening is one of the most important parts of language development. Newborns instantly recognize their parents' voices and can already begin to recognize the difference between their parents' language and other languages. By 18 weeks of pregnancy, your baby can already hear your voice, recognize the tone of your voice, and listen to all the other members of the household.[6]

Reading aloud reinforces language development as your baby listens to you read. When you read aloud, your baby continues to hear *your* voice, pitch, and tone. Your baby recognizes words and sentences and begins to understand patterns in speech. For example, say the following sentence out loud:

I love you, my sweet little baby.

You may notice that you paused at the word *sweet*, and you probably changed your pitch, increasing it at the word *love*, decreasing it at the word *baby*. That shift in tone teaches your baby speech and sentence structure.

You probably did not say the sentence like this:

Iloveyoumysweetlittlebaby

Why? Because you know and understand the flow and pattern of the language you are reading and speaking.

As educators we teach children about patterns in speech and reading fluency by reading aloud to them *and* having them read aloud to us. In any preschool or kindergarten classroom, children are reading aloud, either to a teacher or a peer. When you read aloud, you hear the sentences differently than when you read silently. Often we can recognize changes or errors in our own writing when we read aloud. For example, many researchers and writers today use screen-reading tools in their word processing programs to actually read the document back to them to listen for errors.

Reading aloud, whether from books with simple words or those with more complex sentences, helps your baby understand speech and understand the language (or languages) spoken by your family.

After hearing hundreds of books read aloud, by the time children go to school they can tell the difference between spoken language and the language of books because "listening comprehension feeds reading comprehension."[7]

Some books have phrases, the basic elements of which are repeated on every page—for example, "Goodnight house, Goodnight mouse" in *Goodnight Moon*. Your child is not only listening and learning language but *understanding* what the word "goodnight" means.

This repetition helps babies to absorb language and patterns of speech. Understanding these patterns gives children an advantage when they enter school. It helps them to focus during read aloud sessions in preschool or kindergarten and engages them in exploring more complex stories like those found in chapter books.

By listening from birth, babies continue to become engaged when hearing a familiar voice or a familiar story. It is always exciting for a child to hear a story read aloud at school that was already familiar at home.

 Brainy Baby: I cannot recall how many toddlers or preschoolers have come into my classroom, shouting and jumping with excitement while saying, "I have that book at home!" "We read that at night!" This home-school connection is key to building literacy and vital to your baby's success both at home and in the classroom.

This is where the fun begins. Did baby say *mama, amma, dada,* or *baba* first? Or was it a word from a book that you had been reading aloud since baby was born? Listening is also a reciprocal relationship. Just as your baby listens to you, listen to your baby, and when your baby is ready, after listening to you read aloud every day, you will be ready to hear those beautiful first words. You are never too old to read aloud or to be read to.

Tips and Techniques for Promoting Listening

Keep your baby in the room with you and next to you when having conversations. Many parents have adjacent playrooms or playpens in a separate space. Remember that even when your baby is sleeping, she can hear you!

Just like we play our favorite songs in the car and encourage our children to sing along to classics ('80s music in our household), audiobooks are a great way to continue to build listening skills and language development. Include audiobooks as part of your in-car repertoire.

Quiet time is an important part of developing listening skills. Just sitting together outside and listening to the sounds around you or listening to the sounds in the house as you hold your baby and rock him to sleep will strengthen your baby's ability to focus and listen.

Supporting Attention Span and Memory

Five-year-old Lola comes home from kindergarten and runs straight to the family room to turn on the television. Her favorite show was recording while she was in school. "No, Lulu," her mom says. "You know the rules: no television during the week."

This scenario is common in many households across the country. As parents we want our children to limit their screen time; we put restrictions on when they can watch television, play video games, or watch movies. I wish I were as strong as Lola's mom. Growing up in the 1980s, I was a product of the Nintendo generation, spending hours on weekends trying to beat *Mega Man* or *The Legend of Zelda*.

If I let my three young children play video games for eight hours in a day, that would be judged as bad parenting. Why? Because of the research over the past two decades showing how too much screen time affects children's attention spans. Articles, stories in the news, and guidelines created by the American Academy of Pediatrics inform us that if your child watches too much TV, she is going to have a multitude of problems.

But the truth is, though research has linked screen time with a decrease in attention span and focus, much of it is still inconclusive. That is, though there are some observable indicators of the relationship between screen time and attention span, a decrease in attention span doesn't always occur. Some researchers have found that screen time doesn't have an impact on children's development immediately but may affect attention span in adolescence.[8] (More on this in chapter 12.)

The important thing to know here is that supporting attention span and building memory are teachable skills. Think about the first time you attended a live performance—the thrill of the stage, the lights, the sounds! If it was a rock concert, you were probably standing up and jamming along to the music. But if it was a classical music concert, you were in your seat, sometimes for hours. What helped you to focus in that moment? The experience of the music. The shared, social experience of being in the audience and the engagement with what was in front of you. The reciprocal relationship between the musicians and the audience.

Supporting attention span is important to a child's success in school and is linked to the cognitive ability to focus and concentrate. How do we teach our children how to pay attention at school if we don't support this practice at home? Consider that a kindergartner may be asked to sit and listen to the teacher present a new lesson. It is hard to sit still, in one place, for a long period of time when you are five years old. And not just for five-year-olds—for anyone who isn't accustomed to this practice.

 **Reading Rocks:** You can help your baby develop attention span, focus, and memory by holding him on your lap when you read. Keep the book close so your baby can touch the pages and point to different sections of the illustrations that you want to highlight as you read aloud. Take your baby's hand and touch the images that you are reading about.

By two years of age, your child will be able to recall and connect with the patterns in a simple book that has repetitive, clear
language. When my daughter, Madhavi, was around two years old,
we would read *Goodnight Moon* together every day. Soon I would
turn the page and she could anticipate what was coming next. At
one point she started reading the book aloud to her older brother.
My dad loved to ask her to listen to him read *Hop on Pop*, and
though she sighed and pretended to fuss, she loved to hear him read
the book aloud. He tried to trick her too, to see if she remembered
the text and had memorized the words. And by the time she was
four, she could recite that book from beginning to end.

I'm sure I'm not the only parent who can recite *Hop on Pop* or
Goodnight Moon from the multiple times of reading them aloud.
Just consider, we built the focus and attention span to memorize
that text as we read; imagine what focus it builds for a baby.

As you begin to develop your read aloud habit, think about
the environment in which you are creating this shared experience.
How can you focus on something important if you are not giving
it your full attention?[9] I know many adults who say that they are
"not good with names" and not able to remember faces of people
they meet. One of the strategies often taught in professional development settings is to focus and look at that person directly as you
speak. This is the same strategy to use when helping your baby
develop attention span and memory.

Once your child goes to school, the teacher will try to help her
develop attention span and focus. As the leading organization on
young children's development, the National Association for the
Education of Young Children (NAEYC) outlines age-appropriate
activities for children from birth to age 8.[10] Typically, younger
children are ready to sit and focus for longer periods of time by
the age of six.

As an educator and parent, I have to admit that these are just guidelines to follow and ways to help build your child's attention span. Every child is unique. For example, my oldest son could spend an hour and a half in art class at the age of four but couldn't focus in his preschool classroom during a lesson on animals. My youngest son was so concerned about "getting in trouble" that he was always focused in his classroom, no matter the activity. Children will develop at their own pace.

So how do you build memory? As soon as your child can speak in phrases, some of the first words you'll hear are "read it again." Hearing language from books repeatedly helps children memorize it, just as an actor would reread a script multiple times to memorize lines. Eight-month-olds can remember certain words that are read to them after two weeks of hearing repeated readings.

Reading the same books over and over again may seem an interminable task, but even at birth, babies have been shown to prefer hearing books that were read to them in utero. Researchers gave newborns a choice of hearing their mothers read either a new book or a book read repeatedly before birth. Using a sucking device, babies responded by increased sucking when they heard the familiar book read to them before birth![11] Our babies are listening to us even before they are born. (If only that great skill didn't change when they become teenagers, but that's a whole other book.)

Just remember, as you read aloud, you are helping your baby to build attention span and focus. Each time you hold your baby and read, he hears the gentle tone of your voice and the patterns of language, and keeps his attention focused on this beautiful experience.[12]

Tips and Techniques for Supporting Attention Span and Memory

Find two or three books that you love or that have become favorites in your home. Read these aloud to your baby every day. The more you read and provide repetition, the more your child will build memory. Soon, when reading a book such as *Brown Bear, Brown Bear, What Do You See?*, you can pause and encourage your child to complete the sentence: "Brown bear, brown bear, what do you . . . *see?*"

You don't have to read a book from the beginning or finish one that you have started. If you are reading aloud a longer book to a young child (such as one from the Berenstain Bears series), pause halfway through. Before continuing the next night, ask, "Do you remember what happened to Sister Bear?" or "Why was Mama Bear upset?" This engages your child in focus, recall, and memorization.

Make sure that your baby can hold the book you are reading. Reading aloud is an aural, visual, and tactile experience—that is, your baby will want to touch the pages, turn the pages, and maybe even chew on the pages. Observe how long your baby focuses on one page or image. Lift-the-flap books such as *Where's Spot?* are great for encouraging and supporting memory development. As you lift a flap, reinforce phrases from the text: "Is Spot there? NO! It's a . . ." Pausing for your baby will encourage her to coo or babble what she sees.

Building Imaginative Thinking

Mom: What do you want to be when you grow up?

3-year-old: Ballerina and a firefighter and a unicorn.

One of the greatest things that books give us is the ability to think with our imagination. As we read and fall into the world that the author creates, we mentally experience the space and place, the clothes someone was wearing or the smell of the food being cooked. However, the ability to mentally visualize a world through words is not a skill that we are born with; it is actually a skill that is taught.

Children's books, board books, and picture books do a fantastic job of bridging the narrative (text) and visual (illustrations) to help children build and expand their imaginative thinking.

You may be wondering, Why is imagination so important? Is it something we can teach or a concept we can measure? And why do we lose that sense of imaginative thinking as we get older?

Let's go to one of the individuals who inspired my creative career. The quote that accompanied my senior picture in high school was from Walt Disney. He said, "As long as there is imagination, there will be dreams and the magic of music." Imaginative thinking is rooted in the ability to think creatively, find new answers or ways of solving problems, and take risks. If you recall, these are also the developmental traits of a four-year-old child. If we are born with the innate ability to be imaginative, what happens? Where and when did that level of creative thinking change? In fact, some of the ways that children naturally explore their world build their creative and imaginative skills.[13] This includes every time your baby

- explores new ideas
- tries to solve a problem on his own
- plans and tests ideas and solutions
- learns through trial and error
- take risks (in a safe space), and
- learns to accept failure and try, try, try again.

Over the past two decades as an arts researcher and early childhood educator, I have observed the ways in which our school settings change from preschool through the elementary years.

Preschool classrooms are often set up with centers for free explo-ration. Kindergartners share round tables where they look at each other and interact. By the time your child reaches middle school, she will be in a single desk, facing forward.

These shifting classroom layouts affects how our children learn to socialize, interact, and think creatively. The informal explora-tions of a preschooler shift to standardized tests in middle school. As an arts educator, this is one of the most frustrating aspects of my work. We can all agree that the arts support imaginative thinking and creativity, but what if you don't have access to arts education or a parent who is a musician or artist? What is an accessible way to foster imaginative thinking at home?

You must be wondering why this even matters. In a research study conducted by Fortune 500 companies, CEOs were asked for the number one trait they look for when hiring an employee. Any guesses?

Imagination. The ability to think creatively was valued over hun-dreds of other traits, such as being organized or able to manage others.

One of the ways to support imaginative thinking at home is through books. When you read aloud to your baby, you are not only supporting brain development, attention span, and focus but also helping to build your baby's imagination. There is a reason we say that you can explore the world through books. Imaginative thinking helps your baby to better understand the world around him and to be innovative and creative in life.[14] That doesn't nec-essarily mean your baby will become a unicorn (although, who knows!), but your reading aloud coupled with a child's natural curiosity continues to foster imaginative thinking.

The World Literacy Foundation emphasizes that "reading enhances imagination."[15] Through reading and reading aloud, your baby hears words describing images, settings, and people, and she then creatively pictures the story in her mind. Think about the last time you read a book that was later turned into a movie. When you read that book, you were already creating how the characters looked and interacted, their costumes and settings, the food they ate, and the places they traveled. How did that match with the depiction of that story on film? Your imagination created a version of that book, and by reading aloud to your baby, you are helping him to do the same—to expand his brain by creating the world within a story.

One of my colleagues, a reading specialist for over 40 years, shared some examples of how she would read aloud to her son at home: "Creativity and innovation are so important with supporting social-emotional growth, so as you're reading to your child, find props. We always used props with reading aloud. Let's say you have a story about the kitchen, then read it in the kitchen. If you have a story about leaves and being outdoors, then read the book outside! Play with leaves, let your child touch and feel and crunch the leaves you are reading about. What a kinesthetic, physical experience that enhances your read aloud habit!"

Brainy Baby: When reading aloud, ask your baby questions based on the context of the story. What is going to happen next? Where is the red balloon? How many pieces of dim sum would you eat? What other animals can dance?

In the past few years, there has been an amazing number of children's books that engage and augment imaginative thinking. Here are a few:

- *The Dot* by Peter H. Reynolds (How can one dot change?)
- *Giraffes Can't Dance* by Giles Andreae (Or can they?)
- *Hair Love* by Matthew A. Cherry (Everyone's hair is beautiful.)

- *The Day You Begin* by Jacqueline Woodson (It's the first day of school. How would you handle entering a new environment?)
- *The Day the Crayons Quit* by Drew Daywalt (Not everyone blue wants to paint the sky . . .)

One of my favorite interactive children's books is *Press Here* by Hervé Tullet. (I love this book so much that I mention it again when I discuss how to include art in read alouds in chapter 11.) The simple act of "pressing" into the book creates new and exciting experiences for children. Tullet has many books that follow this pattern of engaging and inviting children to press, touch, and feel a part of the story. This experience builds your baby's imagination, helping her to feel that she is actually creating the story.

I selfishly include some of my own board books here from a series I created called This Is Music, published through Penguin Workshop. Each book invites babies and toddlers to think about how they can create music with anything around them. Can a table be a drum? Yes! Can a paper towel roll be a horn? Of course! What else can you use to make music at home? At the playground? In the car? And these books are interactive. They can actually be played.

Helping your baby to explore the world around him through books will continue to support imaginative thinking. I remember when my youngest, Arjun, was reading a book about the moon with me. He turned to me and said, "I want to ride to the moon on a buffalo." I didn't ask why a buffalo or why the moon. I just asked if I could join in on the ride. Because it seemed perfect.

Reading aloud and asking your baby or toddler questions helps to develop her imagination and ability to think creatively. This simple activity while you read aloud—asking questions that connect with the text—helps your baby to create her own ideas, find solutions, and see the world in different ways.

Tips and Techniques for Building Imaginative Thinking

Board books that have different textures are great for your baby to feel different patterns associated with images and build creative thinking. One of the books Jagan and I used to love to read is called *That's Not My Tiger*, part of a series of board books by Usborne Publishing. These books integrate different textures and patterns for your baby to touch as you turn each page to find the tiger.

Around 6 to 8 months of age, your baby will search for hidden items. For example, if you hold a soft, squishy ball and then quickly tuck it under a blanket, your baby will search for it. This was first documented by the cognitive psychologist Jean Piaget, who did this activity with his own children and labeled this stage of development as "object permanence."[16] To stimulate your baby's imaginative thinking (and cognitive development), after reading your book, ask questions such as "Can you splash like a fish?" and "Can you buzz like a busy bee?" The book *Let's Go Outside!* is part of a series of board books called Indestructibles. With various titles focused on colors, animals, and shapes, these books stimulate your baby's imaginative thinking with simple questions and engaging images. They are also chew proof, rip proof, and washable.

2

BENEFITS OF READING ALOUD FOR LANGUAGE ACQUISITION

———

ONE OF THE MOST IMPORTANT PARTS of child development is teaching language—the ways in which we communicate (spoken and written) in a coherent manner. But how do you teach language? Is it through conversations? Texts? Books?

Think about how we communicate today and how technology has changed in the past 10 years. Picking up a phone and having a conversation has shifted to typing emojis and icons of smiley faces, fruits, and hearts. How did we learn this new language, and how will language have changed when your baby is a teenager?

Even though our forms of communication have shifted greatly, the constants are that we still listen, read, write, and look at images and pictures, with each piece increasing our ability to communicate. Language development is complex. There are two components to language development. The ability to understand language is called "receptive language," and the ability to communicate is called "expressive language."

For example, your baby understands the meaning of your words through pitch, inflection, and your actions (receptive language) before being able to repeat those words or communicate back through spoken language. Imagine that you see a one-year-old who might be trying to climb the fridge (yes, mine did that) and you sharply say "No!" Your baby understands what that word means before being able to communicate what was so exciting about trying to climb the fridge in the first place.

Both forms of language development often occur simultaneously, though that is not always the case. What is most important is recognizing that the critical period for both receptive and expressive language development for your baby is between birth and age 3.[1]

Imagine the following scenario:

A father holds his new baby on his arm. With the other he holds a board book, Spot Loves His Daddy *by Eric Hill, and begins reading each page aloud. The baby has his eyes closed; he is only a few days old.*

What value is there to this read aloud? Even though he was just born, the baby is listening, learning new words, and hearing spoken language, pitch, tone, and patterns. The repetitions in the book reinforce language acquisition and the baby's language development. He might not see the outcome right away, but this father is planting the seeds.

Eighteen months later, the father is reading aloud Spot Loves His Daddy *to his son. This time, the baby looks at the pictures. He reaches out to touch the page and turns it. He may even repeat the words as he listens.*

The repetition of reading aloud shaped this baby's language development. It helped him to hear his father's voice, recognize important sight words, identify patterns in speech, make connections between words and images, and understand what words mean.

It is also important to understand that speech and language are two separate but important parts of your child's development. The Mayo Clinic lists milestones that can help guide your understanding of your baby's speech and language development, and I include them here as a reference guide.[2]

By the end of three months, your child *might* do one or more of these:

- Smile when you appear
- Make cooing sounds
- Recognize your voice
- Cry differently for different needs

By the end of six months, your child *might* do one or more of these:

- Babble and make a variety of sounds
- Use her voice to express pleasure and displeasure
- Move his eyes in the direction of sounds
- Respond to changes in the tone of your voice

By the end of 12 months, your child *might* do one or more of these:

- Try imitating speech sounds
- Say a few words, such as *dada, mama,* and *uh-oh*
- Understand simple instructions, such as "come here"
- Recognize words for common items, such as *shoe*

By the end of 18 months, your child *might* do one or more of these:

- Recognize names of familiar people, objects, and body parts
- Follow simple directions accompanied by gestures
- Say as many as 10 words

By the end of 24 months, your child *might* do one or more of these:

- Ask one- to two-word questions, such as "Go bye-bye?"
- Follow simple commands and understand simple questions
- Speak about 50 or more words
- Speak well enough to be understood at least half the time by you or other primary caregivers

Notice that the word *might* is emphasized in these lists. I repeatedly stress that these are guidelines—while my youngest had over 50 words by 24 months, my oldest had 2 words in his expressive vocabulary by the same age. What is most important is recognizing that these are stages of development, just like being able to hold a pencil or ride a bike: any specific concern or diagnosis should only come from a pediatrician.

What you *can* do is encourage the cooing and babbling when it starts through read alouds, find books that reinforce the words your baby has learned (or learned through books you have been reading), and ask questions. This is called dialoguing. Researchers have found that the dialoguing that occurs between a parent and child during read alouds strengthens the child's language ability and increases connections that occur within the brain.[3,4]

Dialoguing and "dialogic reading" involve reading the same book over and over again with your baby, but expanding on the interactive nature of reading aloud. For example, if I am reading *Goodnight Moon* (a favorite in our household), I would read this book every night and pause to ask questions. These questions ask for simple responses. Ask, "Where is the red balloon?" (your baby may point before verbally replying) and then reinforce the correct answer: "Yes! There is the red balloon." In a video I have of Arjun (around 15 months old at the time), whenever we got to the page of the red balloon, he would start singing the song of the same title, "Where, oh where is the red balloon . . . where, oh where can it be?" This dialogue of reading, singing, and making connections expanded our read aloud experience.

Dialoguing while reading—basically having a dialogue while reading aloud—also has different levels that are expanded based on your baby's language ability and language development. The more specific questions expand to open-ended questions where you interact with your baby about what is happening in the story. This would then expand to include more advanced concepts and ideas that help your baby or toddler make connections with her own daily routines.[5] You might ask, "How do we go to bed at night?" and after she responds, help your toddler make connections from the story to her bedtime routine.

These steps of dialoguing, conversing, and engaging with your baby while reading aloud are the building blocks for reading comprehension skills that he will learn in school. In my experience, most elementary teachers engage in dialoguing during read alouds with their students, pausing during read alouds to ask questions, engage with the class, and help young readers learn to make connections between the story and their own lives.

This example of dialogic reading (taken from the Reading Rockets multimedia project) is a great example of how a teacher would engage her class during a read aloud:

Completion question: "I'll huff, and I'll puff, and I'll

_____."

Answer: Blow your house down.

Recall question: Which house couldn't the Big Bad Wolf blow down?

Answer: *The one made of bricks.*

Open-ended question: Why do you think the first pig built his house out of straw?

Answer: *It was the easiest to build. He was lazy.*

"What" question: What kind of animal was after the pigs?

Answer: *Wolf.*

Distancing question: How do you think the pigs felt when the wolf tried to get them?

Answer: *(Answers will vary.) Scared, angry, sad.*

Home question: If you had to build a playhouse at home, what kind would you build?

Answer: *(Answers will vary.) Tree house, tent, fort.*

School question: The wolf was a bully. He was mean to the three little pigs. What should you do if someone is bullying you at school?

Answer: *(Answers will vary.) Tell a teacher. Tell them to stop. Ignore them.*[6]

As you can see, the teacher begins with a question and expands on the idea with more open-ended questions before helping the students to expand on the story and make connections within their daily lives.

As I have said earlier, every parent is a teacher, and you don't need extensive training or a background in education to dialogue with your baby. Dialoguing and dialogic reading help your baby to understand the story in a deeper way, make connections with her own life, and continue to expand language development.

Read alouds are one of the most important experiences for supporting your baby's language development. Reading to your baby benefits language development before birth and helps him to be ready for that first day of school, laying the groundwork for future academic success.

The Importance of Sight Words

As Sofia is driving her nephew to preschool, he suddenly screams "Stop!" She stops the car and looks back. "¿Que?" she asks, meaning "What?" Antonio repeats "Stop, stop" while smiling. He points out the window, where Sofia sees the stop sign. Antonio has recognized a very important sight word from seeing it every day on the way to preschool.

This simple vignette reveals the key skill of good readers. Do you know what it is?

Many parents might say the answer is being able to read challenging books, being the best reader in the class, or simply knowing phonics, the process of sounding out words.

It would seem that if you know the sounds of the letters in the alphabet, you could sound out and read any word. But sounding out words isn't enough. Without a large vocabulary and an understanding of what you're reading, you're not really reading—you're just saying a blend of sounds. When you already know the word's meaning, it's easy to sound it out, but what good does it do to sound out every word if you don't understand what the words are communicating?

In addition, there are many words that we are unfamiliar with as new readers. Some of these words simply cannot be sounded out. Try this activity:

> **Sound out the word**
> *CAT.*
> "Kuh–Aaaa–Tuh."

That's how we teach phonics in preschool and kindergarten. We show a picture of a cat to match the word.

Can you sound out this word?
THE

This simple word, one of the most used words in the English language, is *not* a word that can be sounded out. Instead it is part of a very extensive list of *sight words*, or high-frequency words, that children need to learn to identify by the time they finish kindergarten. And it is one that really does not have a definition kids can grasp.[7] What picture would help a child define the word *the*?

From the list below, which words do you think are sight words?

am
an
been
did
had
into
must
not
please

Guess what? *All* of these are sight words and are part of a list of 50 to 100 words that children should be able to recognize by the time they are done with kindergarten, and the over 400 they should know by the time they finish second grade. Here is a sample of some of the sight words that your child will be expected to know by the time she is done with first grade:

I	about	all
a	after	am

an	did	help
and	do	her
ask	down	here
at	each	him
ate	eat	his
away	every	how
be	find	if
because	first	into
been	five	is
big	four	it
black	from	jump
blue	funny	just
brown	give	like
but	go	little
by	good	look
can	green	made
came	had	make
come	has	many
could	have	
day	he	

I could keep going.

To a new parent, this might seem overwhelming. How can a baby learn to recognize over 100 words by sight? This is where reading aloud becomes so important. Each time you read a book that has sight words and words that are repeated, your baby makes a connection between the sound of the word, the meaning of the word, and what the word looks like. Research has shown, and continues to show, that the number of words children learn between birth and age 5 isn't in the hundreds but the millions.[8]

As adults we have mastered sight words. Do you sound out every word you read in a book, online, or in a text? No! Because we have mastered sight words and high-frequency words, we do not need to spend time decoding each word that we read. It's a domino effect—once you plant that seed by reading aloud, reading fluency, pattern recognition in language, and vocabulary development expand. The sight words are also grouped as Fry words (many elementary teachers also like to call these "popcorn" words).

It is similar to memorizing the multiplication table in mathematics. Some components of literacy have to be memorized, not sounded out or solved. The International Reading Association recommends that you teach sight words by reading books to young children because by doing so, young children understand these words in the contexts of books. As a result, children begin to seek out texts that are of interest to them.[9]

With all this in mind, there really isn't just *one* skill to identify a good reader. Rather, by supporting a child's ability to decode words *and* recognize high-frequency words, we help the child ultimately begin to find books that are interesting and appealing, thus building a lifelong love of reading. As always, remember that children use different strategies to read aloud, build reading fluency, and develop an understanding of what they are reading.

Some great books that reinforce sight words are:

- *Goodnight Moon* by Margaret Wise Brown
- *Dear Zoo* by Rod Campbell
- *Should I Share My Ice Cream?* by Mo Willems
- *If You Give a Cat a Cupcake* by Laura Numeroff
- *Brown Bear, Brown Bear, What Do You See?* by Bill Martin Jr

You can also refer to any of the books by Dr. Seuss or in the Bob Books series, which all reinforce sight words with high-frequency words and repetition.

Certainly some of these books are geared toward older readers (two- and three-year-olds), but many of these titles are available in different formats. When we think of what types of books to read to a baby, we tend to think of board books as being more accessible and kinesthetic. This is because it is important to have books that (for example) your six-month-old can hold (and maybe even chew) on his own.

Reading Rocks: When your baby is around a year old, you'll start emphasizing the meanings of vocabulary words in books that are part of her daily experiences. As she becomes more social, you will use books to teach daily social interaction phrases, such as "thank you," "hello," and "goodbye." You will be talking as much as reading when reading books that correspond to your baby's experiences, such as going to the zoo, looking at cars and trucks, eating breakfast, and taking a bath.

I cannot stress enough the importance of keeping a list of books while you prepare for your new chapter in parenting (or even if your baby is already born) and creating your own library. In this way you have books to access when you are ready to help build and reinforce these important literacy skills, prepare your child for a lifetime of success, and help his brain grow bigger. Have I said that enough?

Tips and Techniques for Learning Sight Words

The sight words your baby will need to know by the end of kindergarten are all around you. A stop sign is a sight word that you will see every day. As you drive your baby and she becomes more alert to the world around her, show her the stop sign. Say, "Look at the word *stop!*" and then spell the letters: *s-t-o-p*. Repeat this with other words that you see around you often, and soon she will point and say "Stop!" (Or as my kids ask, "Mommy, why didn't you stop at the stop sign?" Sometimes I only roll my stops, I admit!)

Create mini flash cards for your growing baby. Around the age of one, Jagan could point to his name if I drew it on a piece of paper or if it was embellished into his chair. Encourage your baby to learn sentence structure by placing two or three words together. This is also a common strategy used among speech therapists. Using mini flash cards encourages your baby to understand how sentences are formed and to make choices. For example, you could place cards to encourage your child to create sentences such as "I like blue" or "I am two." Though these activities are geared toward your 12-month-old and older, speaking these words and using books to read aloud the words (reinforcing the visual and aural connection of read alouds) will help your baby be ready for reading on his own.

Label, label, label everything. Some researchers suggest that this is a good way to learn unfamiliar or new words (I agree), but labeling the items we use every day as well will reinforce your baby's connection to written words and spoken language. Labeling *on* and *off* on a light switch, *in* and *out* of the bathtub, and common items like *door* and *table* will encourage your baby to connect words she is using in speech to the written word.

Learning Uncommon Words

A study published in 2012 by the *Journal of Speech, Language, and Hearing Research* showed that rare (or uncommon) words were easier for preschoolers to learn than common words.[10] Researchers found that preschoolers who learned rare and uncommon words showed an advantage during comprehension assessments. This is also connected to speech and language development.

I found this study fascinating because my oldest son didn't speak in full sentences until he was almost four. It was the same scenario for my younger brother, and it baffled my parents and speech therapists (back in the 1980s) because he was attentive and intelligent. How is it that one child (me) was speaking in full sentences at a year old and her younger sibling didn't say a word for years?

There are certainly many theories around this: the older sibling talks for the younger sibling, it's genetic, or there are other developmental issues that need to be addressed. We went through all the same concerns with our oldest son, but what was amazing was that although he couldn't say a simple sentence such as "I love you," he could say "Hi, Bharat."

Yes, *Bharat*. It is my husband's name, and one that is so difficult to pronounce that even his colleagues at work alter the pronunciation. We would revel and laugh about this every day. How can a baby say "Bharat" (BHA-raath) but not say the word "love" or "cookie" or "want" or "bye"?

This fascinating exploration of *how* children learn rare and uncommon words has expanded to more research on *why*. Researchers found that children (and adults) process and understand new and unfamiliar words based on what they already know.[11] That means your baby is processing these new words based on what he has already heard you say to him.

Consider this scenario:

You are reading aloud a board book about different types of food. Your baby reaches out as you point to eating an apple, eating a banana, and eating a pizza. You expand the conversation with exciting words and questions: "Yum! Are you hungry? Who wants mumum?"

What other words can you use to expand your baby's vocabulary when talking about eating? Some ideas include *nibble, bite, chew, chomp, munch,* and *crunch.*

You may think your baby doesn't understand these words yet, but she is hearing how these are synonyms for the word *eat.* She is hearing you speak these variations of vocabulary. This expansion of one word, *eat,* helps your baby build vocabulary and understanding of words not often used in books for babies and toddlers.

Whereas sight words are words that your child has to memorize by the end of kindergarten, high-frequency words are often found in various stories and *can* be sounded out or decoded:

> leg
> table
> box
> show

Let's consider the word *paradigm.* Can you sound that out? Not really. The word has too many unfamiliar letter combinations to actually sound out with our typical phonemes. Is it *duh* or *die,* and what is that *gm* sound? Though it probably won't be in a board book for toddlers, *paradigm* is an example of a word that is both uncommon and impossible to sound out with basic phonics. It is also an example of how uncommon words can be important (for vocabulary development) and difficult to learn. Words like *slither, collide, fluffy,* and *darkness* conjure up a magical world that brings joy and wonder to the ears of babies and toddlers. Not that a child might not hear these words occasionally in conversation, but it's the context of the stories in which they occur and the repetition of the telling of the stories that expand a baby's brain connections and reinforces neural (brain cell) pathways. This reinforcement allows children to quickly retrieve these words when needed in future conversation, reading, or writing.

The National Association for the Education of Young Children suggests many ways to encourage teaching your child uncommon and new words. One of my favorites is to bring home a takeout menu from a favorite restaurant.[12] How amazing would it be to show your child "macaroni and cheese" by having him point to it while he eats? Imagine how empowering it would be for him to be able to point it out at a restaurant the next time you visit.

It is important to strike a balance in our work helping to build our child's language development. While many children's books use sight words and repetitive or rhyming words to help teach language development, reading aloud books that are *above* your child's current speech and language will help to introduce uncommon words and build conversations about what these words mean and how you learn their definitions.

Tips and Techniques for Learning Uncommon Words

We love our children. I know a day doesn't go by that I don't tell my children "I love you!" But what are other words we can use and find in books to help expand our children's vocabulary? What are other ways to say "I love you"? Look for books that build on simple language. "I adore you," "I admire you," and "I appreciate you" all use adjectives that help your baby hear and learn new words that mean "love." What are other words you would use to show love? What about in a language other than English? As you continue to speak these words and your baby becomes closer to a year of age, write the words down in a sentence and show it to her every day. She will soon learn new and uncommon words (not just *love*), and not just in English.

Uncommon and new words include names and the names of corporations. My daughter knew how to spell *McDonald's* because of how many times we passed by a sign. My youngest could say and spell *Starbucks* by the time he was three. Why is this? Repetition and recognition. Yes, we eat at McDonalds's and we love a Starbucks vanilla Frappuccino, but these were also uncommon and new words that my kids learned by connecting their daily experiences to written words. What is your favorite restaurant, and does your baby know the name or get excited when you pull into the parking lot?

Supporting the Acquisition of Different Languages

*As the kindergartners enter the classroom, Ms. Winters greets each of her students with a high-five, fist bump, wave, or hug. As the year progresses, she asks each of them to say hello in a language that is **not** English. She has shared these in class. By November these are some of the replies:*

> *Konnichiwa*
> *Namaste*
> *Bonjour*
> *Guten Tag*
> *Hola*
> *Buenos dias*

This informal way of engaging children in dialogue in a language other than English is a way to invite them to hear and recognize the sounds of other languages (and the fact that they all have ways of saying "hello" or "good morning"). It also encourages children to find and explore the languages of their home environment, their friends in other settings, and their grandparents. All of which leads me to this question: How many languages are spoken in your home? You may only speak English, but your spouse or significant other might be fluent in English and French. How many languages is your child hearing?

In the past decade, families around the country have grown from speaking only English to being multilingual. Not only do the parents speak multiple languages but so do their children. This has greatly shifted the definition of "bilingual" homes. In our home alone, my husband speaks Hindi, Punjabi, and Burmese, while I am fluent in German and conversational in Spanish, Italian, French, and Cantonese. We both speak fluent English and Tamil. I always wondered, How does this cacophony of languages affect our three children's language development?

When Jagan was diagnosed with a severe speech delay, many individuals told us that it was because there were so many languages at home—the variations were confusing for him, he couldn't grasp one particular language, and so he wasn't speaking at all. He was two years old, and the speech and occupational therapists said his language development was below that of a one-year-old.

How was this possible? We read to him every day—every single day. We talked with him. Sang with him. Held him. After all, he was our firstborn. He was everything. As first-time parents we wondered, What did we do wrong?

But I found it hard to believe that multiple languages could restrict a child's speech and language development, and the research I studied supported my theories. After all, how are children in so many settings growing up hearing multiple languages and learning to speak and understand each one? For example, how does a child who lives in a household where Spanish is the primary (or only spoken) language come to school and speak perfect English with his peers?

Basically, the more words you learn, the more words you can speak. If you see or hear the word *blue* in English and you also speak French, you can relate *blue* to *bleu* in French. The brain connections have already been developed using the home language. Concepts learned in one language easily transfer to another language. If your baby didn't have the necessary concepts, she would be held back intellectually. Think of the number of things we explain to our babies in the first years of life that contribute to the development of brain connections. It doesn't matter from which language the brain connections come and expand.

Ultimately, it is best to speak to and read to your child in the language in which you feel most comfortable, even if it is not the language taught in schools. Your child needs to hear a lot of vocabulary and a lot of spoken dialogue, and be introduced to those common and uncommon words. It's better if your child hears lots of all kinds of Spanish words in their context rather than fewer words in English.

There is also some evidence that children with a well-developed home language will successfully learn other languages and do well in school regardless of what language is spoken in school. The connection here is that children who have not heard enough language, no matter what that language is, usually struggle learning to read.

It is not *which* language that matters but the amount and quality of the language children hear between birth and 3 years of age. If children's home language is filled with rich descriptions from books and detailed explanations from parents, they will have more language to be able to question, investigate, and make sense of their environment. The more language children hear, the more intelligent they become. This is a snowball effect—when a child has a strong grasp of language, he will become a strong communicator and more successful in reading, math, science, and the arts.

If English is not your first language, participate as a family in English-language community events, such as weekly library read alouds or playgroups, sports events, and activities at your place of worship. Try to enroll your toddler a few days a week in a preschool where the teachers and other children speak English. Your child's strong Spanish, French, Hindi, or any other language foundation along with her early experiences in the English-speaking community will help her become bilingual and succeed in school. I strongly believe there is no reason to give up one language for another.

Consider some of these guidelines regarding language development. Within the first year of life, your baby will probably:

- Make sweet cooing and gurgling sounds at you
- Blow raspberries
- Smile and laugh
- Turn to face familiar sounds
- Start to formulate words with specific consonants such as *p*, *m*, and *b*

However, I cannot emphasize enough how much these are just traditional guidelines—that is, based on how language development is measured across a large population of children. Does every baby speak like this by the first year of age?

No.

What we have learned through our experiences as parents is that these are only guidelines to observe your baby's language development. A delay doesn't necessarily mean a problem, nor does it mean you made a mistake as a parent. Your child may be way ahead of these stages as well. I know one of my friends said her son was saying "hi" at three months old. I didn't see it, but I believe her.

I wish I had more opportunities to engage with other parents and families who raised children in a multilingual household. Were the differences really so strong? Was it necessary to stop speaking Tamil in order for Jagan to learn English?

After my parenting experiences with three children who demonstrated completely different speech and language development, I learned that the only way a child will learn different languages is by hearing them and ultimately communicating back in them.

Look at this question-and-answer session from a bilingual family about how to support language development for languages other than English:

> **Q:** I'm Chinese and my husband is Canadian. I sing nursery rhymes from my childhood to my three-month-old. I now want to start reading to my baby in Chinese. My husband, who speaks English, thinks hearing two languages will confuse our baby. Is this true, and if so, should we speak to him only in English?
>
> **A:** Babies under six months of age can hear the sounds of all different languages, including some sounds that later become increasingly difficult to pronounce, such as the French *r* for English speakers and the English *r* for Chinese speakers. There is a window of opportunity in the first year to perceive and maintain these and many other sounds. If you want your baby eventually to be able to speak Chinese *without* an English accent, it is important that you continue to sing, read, and talk to her in Chinese. She will also pick up Chinese word order (grammar and syntax) from hearing it spoken.

Will your baby be confused by also hearing English from your husband? The short answer is no. Babies' brains are very pliable and absorbent. Decades ago researchers used to think that young children came into school with a "blank slate," but through more brain research we now understand that from birth through age 3, children are natural "sponges," soaking up everything in their environment.

 Brainy Baby: Your baby can pick up more than one language at once. However, when he first starts talking, your child may interchange some words of both languages because he isn't sure which words go with which language. But as he grows older, he will become aware of who is speaking which language, and he will figure out how to separate them. It is incredibly fascinating to observe! By the time your child reaches school age, he'll have figured out in which situations to use each language and with which people.

Make sure that once your child is in school you continue to read and talk to her in Chinese even if English is the primary language spoken at home. She needs to learn the value of bilingualism in order to speak to her relatives too. If you can travel to China to visit relatives or maintain relationships with the Chinese community in your area, your child will see the value in knowing more than one language.

If you are living in a predominantly English-speaking country (for example), your child will learn English as soon as he becomes part of the larger community. Learning English will be easiest for those children who have the largest vocabularies in their home language, and reading aloud is a great way to help young children learn English. Through reading aloud, children will hear the inflections we use with our voice when we speak the English language, including intonation, rhythm, and pronunciation.[13]

Young children have the incredible ability to easily learn new vocabulary. They also have the motivation to learn any new language in a new setting because they want to become part of the social group and are not shy about speaking a new language.

This is also true for children with speech delays. Jagan didn't talk in full sentences until he was fully immersed in a school setting where the social dynamics forced him to learn how to say, "I want to play with Thomas the train and not James." Even if his pronunciation wasn't perfect, the other children understood.

To make the transition to kindergarten easier for your child, expose her to English by associating as much as possible with English speakers outside of the home in English-speaking community activities and by enrolling her in English-speaking preschool. As a teacher, I have worked with children whose home language is not English and who have higher language ability in English than native English speakers who have not been spoken to and read to enough at home. These children do better in school than their native English-speaking counterparts.

Consider this scenario as well:

Q: My husband is Japanese, and I am a native English speaker. We want our child to know both English and Japanese and grow up to be bilingual. My Japanese isn't fluent, and my husband's English isn't fluent, but we communicate in English. What should we do?

A: Because your husband is a fluent Japanese speaker, he should speak and read to your baby in Japanese, and you, as a fluent English speaker, should speak and read to your baby in English. It is important that your baby hears both languages from the moment of birth. At birth, babies' brains are capable of learning any language. For the first eight months, babies focus on the *rhythm* of the languages they are exposed to. At eight months, babies begin to focus on the *sounds* of their own home language (or languages) and lose the ability to hear and produce sounds in other languages. Hearing the rhythm and sounds of your home languages before age one will help ensure that your baby will be able to speak these languages.

Here's the thing. I am not saying that your child will never be able to learn another language, but most of us have experienced how difficult it is to learn another language as young or older adults.

In addition, your baby needs to continue hearing and interacting in both languages in order to develop and maintain those languages. Your husband could talk and read to your baby in Japanese, and you could talk and read in English. Once your baby goes to preschool, he will hear more English than Japanese. As a result, your husband and any relatives who speak Japanese will need to continue to interact in that language, or English will take over. Your child will hear social language every day from interacting with your husband and his family. However, if your child is going to maintain Japanese after he has started school (and is immersed in English), your husband will have to make a conscious effort to read to him in Japanese to enlarge his vocabulary with more uncommon and academic vocabulary.

On a more personal note, when our second (Madhavi) and third (Arjun) children were born, my husband and I spoke as we always did—in a mix of English and Tamil (lovingly referred to as "Thenglish" among Tamil-speaking Americans, in a manner similar to "Spanglish"). Today all three of our children have a strong receptive understanding of Tamil and can speak to us using certain phrases that we have probably used more often than we realize. For example, "*adha thodaadhe*" is a common phrase I used with my young children when there was something I didn't want them to touch if we were out in public. Needless to say, Arjun finds ways to use that phrase with me all the time, such as when I am going to watch a show on my iPad and he wants my attention instead.

He will say, "Mom, *adha thodaadhe*."

Point taken.

Not every family will have multiple languages spoken at home. But recent work to increase the diversity and inclusion in children's literature has brought a push for language equity that has made it easier for families to build and support the acquisition of multiple languages at home.

Here are some great books that have been printed in both English and Spanish:

- *Besos for Baby: A Little Book of Kisses* by Jen Arena
- *Animal Talk: Mexican Folk Art Animal Sounds in English and Spanish* by Cynthia Weill
- *Buenos Noches, Gorila* by Peggy Rathmann (*Goodnight, Gorilla*)
- *Los Pollitos* by Susie Jaramillo (*Little Chickies*)
- *La Oruga Muy Hambrienta* by Eric Carle (*The Very Hungry Caterpillar*)

The wonderful experience of having books printed in dual languages is that while you may read the book to your baby in English, Abuela might read it the next night in Spanish.

But speaking the language at home isn't the only way to support your baby's acquisition of different or multiple languages. There are also musical companies that record songs and lullabies for babies that are inclusive to cultures all around the world. Playing these soothing pieces at nighttime, after reading, or during quiet playtime will help your baby to understand the inflections of different languages even if you don't speak them at home (more on this in chapter 10, when we discuss using music to support read alouds).

Tips and Techniques for Supporting Acquisition of Different Languages

Based on what languages are spoken at home, either by each parent or other relatives, build a library of books to support that specific non-English language development. Two decades ago I was surprised to find a shelf of "international books" in our local library in Illinois. How was there a book written completely in Tamil, and as a board book? The acceptance and recognition of the languages we all speak has expanded to books being published in dual languages. A wonderful and accessible way for you to support your baby's bilingual (or multilingual) development is by having the same books in two different languages.

Numerous apps not only support the teaching of multiple languages but also encourage the acquisition of different languages through word games. This would be a perfect intergenerational activity for your preschooler and grandparent to connect through your native language.

Create your own book at home with simple sentences written in a language that you speak at home. This is a simple project that will become a beloved keepsake as your baby gets older and can write her own sentences to add to the book.

3

BENEFITS OF READING ALOUD FOR SOCIAL-EMOTIONAL HEALTH

Preschooler:	I'm scared to go to school.
Mom:	Why? School is so much fun!
Preschooler:	(*Shrugs*)
Mom:	What's scary?
Preschooler:	What if I throw up in my mask again and everyone makes fun of me?
Mom:	Everyone throws up. It's OK.
Preschooler:	What if my teacher is mean?
Mom:	(*Sighs*) . . . (*To be continued . . .*)

This was a real conversation between myself and my sweet little Arjun in 2019, right before the pandemic started, but since then, this type of conversation has resonated with parents all around the world.

Sure, young children are always scared of school, scared of new social experiences. But something changed in the past couple years. Our children are growing up in a world where we were masked and really unknown. I look at my niece, who was two when the pandemic hit (she's now four), and wonder how she will interpret and understand social cues, emotions, and feelings from her peers (and adults) when those around her were always masked?

I think about my own challenges with balancing my anxiety and apprehension about my three children's well-being with motivating them to be excited about returning to school feeling prepared. As parents we found coping strategies for our own social-emotional health and well-being. Many of us exercised more, many of us ate more. While many of us didn't know what day of the week it was, as adults we have years of strategies to support and maintain an appropriate level of social-emotional health, because we have the capacity to acknowledge, cry out, or yell for help.

In the past two years, the strongest area and interest in early childhood research has been in understanding and supporting our children's social-emotional development. Social-emotional development (also discussed as social-emotional health and well-being) is defined as "infant mental health" within the field of early childhood research.[1] At a very basic level, social-emotional health is a baby's ability to explore and regulate emotions and build strong, meaningful relationships.

Think about your baby's first bonding experience. It is with a parent or caregiver. From birth a baby is learning to understand what a relationship means, how people respond and react to certain behaviors and cries, and what it means to be held and loved. If a baby doesn't have an established understanding in the home environment of what it means to respond to emotions, regulate one's own emotions, or sustain a meaningful relationship, those absent definitions and experiences will directly translate into the school environment.

Despite decades of ongoing studies focusing on how to build social-emotional competence and support young children, researchers, educators, and parents are really still asking the same questions: How do we support our children's social interactions and understanding of one another? How can we strengthen our children's emotional intelligence and health?

Let's break this down even further. For young children, social competence is the ability to make and sustain social connections through skills such as being cooperative and flexible. Think of this as the very common toddler and preschooler situation in which "I want that toy" *should* become "Let's take turns." Is that always the case? Probably not.

Emotional competence, on the other hand, is the ability for a child to self-regulate his own emotions and understand and interpret his peers' emotional cues. Think of this as a situation in which a young child in a playgroup falls and gets hurt. Do any of the other children stop to ask the young child if she is OK? To ask if she is hurt? To help her up?

Emotional competence is also about expressing and understanding emotions. This is one of the most difficult skills to teach. Consider ourselves as adults. Are we happy all the time? We have our good days and bad days. Why then, as parents, do we often expect our children to be happy, flexible, outgoing, and easygoing *all the time*?

We can teach social-emotional skills and competence, and we should begin as soon as we are reading aloud. Think about how children will start crying if another child cries. They aren't copying that behavior; they are reacting to the emotion and sound. If my oldest, Jagan, started crying (when he was three), it was expected that my daughter, Madhavi (one year old at the time), and my youngest, Arjun (a newborn at the time), would all join in the fun. Even if it wasn't the best parenting moment, at least I knew they all loved each other (or at least reacted to each other's wails).

Social-emotional skills and development begin at birth, and the ways that you interact with your baby help to foster these skills. When infants' needs (such as being held and fed) are consistently met by adults, they are better able to regulate their emotions, pay attention to their surroundings, and develop strong bonds and relationships.[2]

Within your baby's first year of life, he becomes more aware when he is separated from an adult caregiver, learns to be comforted and soothed, begins to participate in social interactions and communications through body movements and smiling, responds to his name, and starts to vocalize with you. During this time your baby is also developing the ability to express how he feels, including sadness, happiness, fear, and anxiety.[3]

Reading Rocks: You can help foster social-emotional development through read alouds with your baby. Every time you snuggle together with a book and she hears your voice, responds through movements and vocalizations, and listens to your pitch and inflections, you are supporting your baby's social-emotional skills.

Social-emotional development isn't just about helping your baby understand and interpret social cues or gain empathy toward his siblings or peers. Research has shown that bullying and aggressive behaviors begin as early as preschool.[4] I observed this with my oldest. When Jagan entered preschool at the age of four, his speech was still not fully developed, and so he would be pushed and shoved around in the playground until a teacher intervened. He simply couldn't use his words to express how he felt to his peers.

What has been challenging for schools and communities to understand is how social-emotional competencies do not necessarily fit in a box. Growing up in India, my mom could have never looked at an adult in the eyes or run to help a peer if a teacher told her to stay in her seat. There are cultural norms that we have to acknowledge as well, and books are a great way to develop not only your baby's understanding of social-emotional skills but also your own. There are two pieces here: the cultural aspect and the societal aspect of social-emotional development. The best we can do is acknowledge what our children are going through and find books to help them talk about and support those feelings. Some great books that I have used with young readers include the following:

- *The Big Boy with Big, Big Feelings* by Britney Winn Lee
- *Ruby Finds a Worry* by Tom Percival
- *Little Monkey Calms Down* by Michael Dahl

I also really love the book *Making Faces: A First Book of Emotions* published by Abrams Appleseed. This book and other board books that have images of babies' faces or mirrors for your own baby to look at are a great start for exploring and understanding social-emotional development.

All of this matters because research shows that children who have healthy social-emotional development are more confident, take greater risks, and build stronger relationships. Specifically, social and emotionally healthy children demonstrate the following behaviors and skills:[5]

- Are usually in a positive mood
- Listen and follow directions

- Have close relationships with caregivers and peers
- Care about friends and show interest in others
- Recognize, label, and manage their own emotions
- Understand others' emotions and show empathy
- Express wishes and preferences clearly
- Gain access to ongoing play and group activities
- Are able to play, negotiate, and compromise with others

Research has also shown that stronger social-emotional development early in life is related to better academic achievement and helps your child build positive relationships with others.[6]

Supporting your baby's social-emotional development has become increasingly important over the past few years and is also an integral part of curriculum in school. The fact is that by reading books aloud to your baby, you will help to build a bond and connection, foster your baby's understanding of empathy and respect, and help to create a lifelong love of books.

Creating a Bond and Connection

Have you ever heard of the word *parentese*?

To be completely transparent, I never knew that the singsong, lovey-dovey, high-pitched way that I spoke to all three of my kids actually had a specific term. Parentese is an important and necessary form of communication with your baby. When you hold your baby, you will find yourself talking in a higher-pitched, musical voice that is also slower and somewhat louder than traditional dialogue. When reading you'll use the book primarily as a vehicle to converse with your baby using your parentese voice. Studies show that beginning at around five weeks, babies actually prefer parentese to adult conversations (or the tones we use to speak to other adults). In fact, new research in 2020 actually encouraged the use of parentese to support babies' speech development.[7]

Why? Because speaking parentese creates an intimate connection between you and your baby. This is also the strongest way to support your baby's social-emotional development through speech and read alouds.

Some of the main features of parentese are as follows:

- Put face very close to the baby's face
- Use a higher pitch in the voice
- Speak in a melodious tone (almost like a song)
- Articulate clearly
- Frequently use repetition
- Use exaggerated facial expressions (eye contact, raising of eyebrows, and big smiles)
- Move body rhythmically
- Lengthen vowels ("soooooo cuuuuuute")
- Use shorter sentences
- Always sounds happy

Parents all over the world speak an intimate form of parentese. Isn't it amazing that regardless of language, the social-emotional way in which we communicate with babies is nearly identical in our tone and pitch?

But here's the thing. Parentese is not baby talk. This isn't holding your baby and saying, "Goo goo ga ga boo boo baby!" Parentese is how you change the inflections in your tone and voice to almost mimic a musical melody. You would probably never hear a father speak to his newborn in a monotone (or robotic pitch without inflection). Even if you don't realize it, you are speaking parentese.

Yes, we also engage in baby talk. For example, researchers have also distinguished between baby talk and parentese. Baby talk is the actual altering of the spelling of the words to utterances bordering on what might be perceived as nonsense. It can turn a sentence like "Look at the cute little baby" into "Wook at zu coot wittle bubu."

Saying something like "Oooh, your shozie woozie on your widdle feets" is considered baby talk and not parentese.[8]

Though parentese might feel and sound simply cute, the reality is that your baby is also processing this as conversational speech. Some parents may even feel uncomfortable speaking parentese because they are not used to using it, or they may think their newborns can't understand, so why bother?

Researchers have found that parentese, as a more natural, instinctive form of social-emotional communication, is important for your baby's speech and language development. It helps to build a bond and connection and should be encouraged when reading aloud. In fact, researchers also explain that reading aloud is a shared experience for building connections with your baby. Over 90 percent of parents noted that the experience of reading aloud with their baby was a positive, loving experience that helped them to feel closer to their baby.[9]

Children who have been read to remember this experience through their teenage years, recalling how it was quality time and helped to create special memories. In fact, when teenagers recall their fondest childhood memories, they most often say that they were read to at home.[10]

Think about your own childhood. Do you have memories of being read to? Or did you read to a sibling, cousin, or friend?

And reading aloud doesn't just have to be reading aloud to humans. My daughter, Madhavi, sits and reads books aloud to our dogs. And the dogs love it. Our terrier, Brownie, actually fell asleep during one of her read aloud sessions while my son Jagan read aloud to his bird, Beethoven, who perched intently in his cage and even whistled appropriately. It was lovely to watch. These experiences reinforce the importance of reading aloud for my children while also helping to create a bond with very special members of the family.

I don't know if the dog and bird are building their vocabulary from these read alouds, but what it's teaching my kids is that read alouds are a reciprocal experience: reading aloud to someone (or some pet) is different from being read to, and that understanding is valuable.

Reading aloud is one of the easiest and least complicated of all the daily tasks that you do with your baby. It helps you bond and attach to your baby because you're putting everything aside to give full attention to your precious little one. It promotes family togetherness. It's a built-in calming routine for you and your children.

Tips and Techniques for Creating a Bond and Connection

Along with parentese, the other two main features of reading to babies are dialogue (or conversing) and questioning. For newborns dialogue occurs when they respond to read alouds by moving their hands or feet, cooing, breathing faster, or giving you some bodily signal of response and pleasure. After you talk and read a little, allow your baby time to respond in some of these ways. Then respond by talking some more. For example, when your baby starts moving her arms or legs as you read and look at a page with a bright yellow duck, you might say in your singsong voice, "You like that page because it's about that cute little yellow duck, like the one you have." This is a natural way to build connections with what you are reading about and what your baby is experiencing every day. Trying to read the book without responding to your baby will feel unnatural. Try it both ways, and you'll see for yourself.

Find different places in your home environment (or even outside) to read aloud. When I was pregnant with my second child and too exhausted to run after my first (and boy, could he run fast!), I would pull up a blanket in the backyard and lay out books. Together Jagan and I would flip through the books. Sometimes I would read aloud a specific book, and other times we just explored the stories together. He still remembers this bonding experience today. These different spaces will create different experiences and memories for reading aloud.

Remember that you're also teaching your baby the reciprocal experience of communication while giving him a constant stream of words necessary for language development. It's also so much more. It's an intimate bonding through words that uniquely happens in the act of reading, talking, and cuddling.

Fostering Empathy and Respect

In 2019 Merriam-Webster chose the word *they* as their word of the year.

This decision marked both a unique and necessary acknowledgment of our times, the ways in which we use language, and how we build empathy and respect for those around us. This was a step that recognized the importance of *they* as a "widely used and necessary component" of inclusion.[11] It also fueled a strong push in books and other media to find ways that accurately support all individuals and help young children to be seen and represented in books.

Why was that decision so important? As conversations around gender and identity have shifted over the past few years, the inclusion of multiple pronouns is one of the greatest movements forward. Not everyone is a he or she or identifies as such, even though the English language (and our conversational language) has always presented gender as this binary choice. Individuals around the world are using multiple platforms to have these discussions, encourage one another to ask rather than assume, engage rather than reject, and build empathy and respect for another person's situation and feelings. Importantly, when children participate in read alouds, they build empathy and understanding of other people's lives and situations, expanded by follow-up discussions and dialogue with an adult.[12]

One prominent children's book that was published on this topic is *What Are Your Words?* by Katherine Locke. This book tells the story of a child named Ari who identifies as gender fluid and nonbinary and explores what it feels like to use and identify with specific pronouns. The book was already received in a strong light by communities across the country for highlighting the challenges of individuals who are trans and nonbinary, particularly children, who are at greater risk for mental illness and depression.[13] The author acknowledged that growing up, the only pronouns that were ever discussed were *he/him* and *she/her*, and while it was freeing to be able to identify as *they/them* as an adult, the conversations should begin with young children, who are more open and accepting of these discussions.

Over the past few years, while living in California and later moving to Chicago, what I observed was that the conversation about pronouns is very polarizing among communities, both liberal and conversative, in large cities and small towns alike. I have encountered adults getting into arguments and fights over why they should even bother asking someone's pronouns, teaching artists frustrated that they have to now "memorize a whole new set of words to teach a class" (someone really said that in a broader conversation about teaching and empathy), and those cheering this new direction of understanding and acceptance. Regardless of where you stand or how you feel about this discussion, we can all agree that very young children learn empathy and respect from modeling at home and in school.

Though gender and identity are an important part of the conversation about empathy and respect, young children also need to be involved in discussions about race, culture, and diversity. It would be irresponsible to not acknowledge the challenges many Black and Brown communities have faced with discrimination and fear, particularly in the past few years. Books such as *Stuntboy, in the Meantime* by Jason Reynolds and *Black Boy Joy* edited by Kwame Mbalia are important for young children to recognize why certain movements and narratives are meaningful and to see themselves in a story that encourages their peers to feel empathy and respect for their stories.

For example, we can talk with our young children about the Black Lives Matter movement, but what does that mean to them? How can they understand the power of this movement and build empathy and respect for one another? One of the ways we can do this is through read alouds, using a story that has a strong message and having a conversation with our children.

Now you may be thinking an infant is not going to comprehend any of these messages, but an infant can recognize the images and colors used in a book. A toddler will cheer at seeing someone who looks like her on the cover. A preschooler will begin to understand the emotions of his peers through books that discuss fear, isolation, happiness, and celebration.

Tips and Techniques for Fostering Empathy and Respect

When I was teaching preschool, and later kindergarten, I always had a feelings space, where children would pick or point to the way they were feeling that morning. These simple images of a smile, frown, fear, sadness, and so on helped children to express how they were feeling and understand how their peers were feeling as well. One of the greatest moments came when a child would ask her peer, "Why you sad?" It would be the beginning of a conversation and relationship with empathy. You may think an infant or baby is too little to understand empathy, but talking about feelings is always a start. As you pick up your crying baby to hold him, you may find yourself automatically responding with words that teach empathy: "Don't cry, sweet pea, Mama's here." As your baby gets older and starts to connect words with language, building his understanding of empathy not just within the home environment but also in school, some of these books are a great place to start the conversation:

- *The Boy & the Bindi* by Vivek Shraya and Rajni Perera
- *Bunnybear* by Andrea J. Loney and Carmen Saldaña
- *When Aidan Became a Brother* by Kyle Lukoff and Kaylani Juanita
- *Last Stop on Market Street* by Matt de la Peña and Christian Robinson
- *Those Shoes* by Maribeth Boelts and Noah Z. Jones
- *Red: A Crayon's Story* by Michael Hall
- *Fry Bread* by Kevin Noble Maillard
- *Adrian Simcox Does NOT Have a Horse* by Marcy Campbell and Corinna Luyken
- *Just Ask!* by Sonia Sotomayor and Rafael López
- *I Am Enough* by Grace Byers and Keturah A. Bobo
- *I Am Human* by Susan Verde and Peter H. Reynolds
- *We're All Wonders* by R. J. Palacio
- *I Am Perfectly Designed* by Karamo Brown, Jason "Rachel" Brown, and Anoosha Syed
- *Eyes That Kiss in the Corners* by Joanna Ho and Dung Ho

These picture books are just a few that address gender and identity, access and equity, and diversity and affirmations. Helping your child recognize that she is valued and cared for will support her social-emotional development, empathy, and respect for others.

Building a Lifelong Love of Books

I remember when I was 17. The last thing I wanted to do was be around or close to my parents. I was a teenager, and my parents were not reading aloud books to me. But the impact of my mom's read aloud habit (every night) stayed with me throughout my teenage years and into adulthood. In a 2019 report published by Scholastic, teenagers reflected on how reading aloud with their parents created "special memories" and that these were "quality one-on-one times" with parents. It was fascinating for me to see how teenagers reflected on how the experience of being read to influenced their own relationships with their parents. As one 17-year-old said in the report, "Reading aloud helped bring me closer to my parents."[14]

Consider this statistic: over 80 percent of children between the ages of 6 and 14 said that they loved being read to by a parent.[15] Those numbers and these informal reflections are enough to validate how important this experience is for building social connections, emotional strength, and a lifelong love of books.

The reality is that even if we are reading aloud during pregnancy, after birth, and with our toddlers, there is a significant decline in how often we read aloud with our children as they get older. I see this in my own family dynamic. I read aloud more often with my youngest than I do with my middle schooler. Why is this? I wonder if it is because of the books my older child may choose to read and a belief that he should have that experience of reading by himself, experiencing the stories and characters in his own world. And why would I feel more inclined to read about dinosaurs with my youngest? Or stories about how to care for horses and dogs and cats with my middle child? When I tell them to turn off devices and "go read instead," am I really supporting a love of books or making it a chore?

Research has shown that about six in ten kids find that reading (either on their own or being read to) is enjoyable. That is 60 percent of children. As one teenager noted, she "didn't have time to read" anymore.[16] Is that the case? Or is reading just not a priority. And how do we build a love of informal reading in a world of timed tests and comprehension assessments?

Recent studies have shown that between the ages of eight and nine, there is a 20 percent decrease in individual interest in reading. That means that while your eight-year-old would read at least five days a week, by the time he turns nine it is probably one or two days.[17]

However, we can continue to build a love of books after we start reading aloud, encouraging reading aloud time, and then hopefully inspiring our children to read aloud on their own (and to others). Here are a few ideas:

- **Kids want funny books.** Kids like things that make them laugh, such as books about dogs who are police officers, underwear, and silly accidents or mishaps.
- **Kids like books about adventure and disasters.** Stories about finding ways to survive and escape featuring their favorite characters help children to connect with conflicts and situations.
- **Kids like books with pictures.** I grew up in a time when comics were not considered books, and though I also grew up in a household where I was read to every day and would spend hours in the library, my school didn't consider my X-Men or Archie comics as "true" reading. This has changed immensely in the past decade— books that were once dismissed are now rebranded as graphic novels.
- **Kids like books they can relate to—or aspire to—in difficult times.** Kids who read more frequently share that they want characters who are smart and strong, with over 60 percent saying that a book has "helped them through a difficult time."[18] That experience will stay with your child forever, helping to foster a love of books.

When a one-year-old independently goes to pull out her books from her bookshelf, she does so because she has experienced the pleasure books can give, even when she just sits for a short time by herself and looks at them. That independent pleasure is associated with the pleasure she gets cuddling with her parents and reading before naps or bedtime, after lunch, or first thing in the morning. The association of books with love and comfort will last a lifetime.

Enjoying reading is an important part of building literacy skills because it involves the emotions *and* the motivation to read. Without the motivation to read, children don't read. As children progress through school, the amount of reading for pleasure accomplished outside of school adds hundreds of words to their vocabulary, many of which are rare words. But with the demands of school and testing, your child has to feel motivated from within to want to keep reading.

Importantly, building a love of reading begins at home (recall how I believe that every parent is a child's first teacher). However, for many children, their first experience of being read to might be in a school setting, in a group read aloud with the teacher. How do we expand this experience?

Building on your child's interests is the first place to start.[19] My daughter never loved reading about unicorns and dolls, but she loved any books with kittens. My oldest son loved books with trains and the LEGO series for older readers, whereas my youngest would listen to his older brother read the entire Who Would Win? series as they compared which animal might actually win the fictional battle. The videos I have of this are charming.

If I ask Jagan (currently a seventh-grader) to sit with me and listen to me read, or to read to me, I know he will say yes. Just last night, my daughter, Madhavi, came to my room and asked if she could read aloud a chapter from a new book she checked out from the school library. As a side note, I was exhausted and may have lightly drifted to sleep at times, but she read aloud for 30 minutes, laughing at moments that were joyful in the book and cherishing this quiet time with me.

And from her read aloud, I learned some new things about how to groom dogs.

I grew up in a home environment where we didn't have much. What we did have was access to the public library. I will continue to argue for how important those library visits were for myself and my younger brother. Not only did books open our world to new experiences but we also had the opportunity to participate in contests, sit in on read aloud sessions, and be read to by books on tape (more on those amazing products later on).

I acknowledge that my mother's lifelong love of books (passed on to her from her father's lifelong love of books) was then an immersive part of our childhood. But it hasn't always been easy with this tech-savvy generation of young children. I have tried to find ways to encourage my children to get off devices and under-stand the acronyms with which they speak: "MOM! OMG, I'm AFK and LOL and IDK if you want me to keep reading."

What?

Finally, after much translation, I started to invite my three kids to spend 30 minutes a day to just read. Read together, read to each other, or read alone. I also asked them to read all my children's book manuscripts before I sent them to my agent or editor. Far before the book would ever appear in the store, my kids had read the story aloud to me or heard me read it to them. (It is also, secretly, a great editing technique. You won't believe the number of times I have missed a simple word in a sentence, and reading aloud helped me too.)

This experience of reading my books, seeing them on my computer or printed on paper, helps to foster my children's love of books as they anticipate seeing them in print. It also reinforces the importance of reading at home, reading aloud to each other, and reading together.

Though you may not be an author, you can still write a story and have your child read it aloud, or vice versa. Read your child's story aloud when he is of school age and so excited to share his new ideas with you.

How wonderful it is that the simple act of daily read alouds leads to so many benefits for our babies. You don't have to be super mom or dad (I don't even know what that means anymore, to be honest). You don't need to be a reading specialist either.

All you need is books.

Tips and Techniques for Building a Lifelong Love of Books

Encourage your child to find books that appeal to her interests. Not all boys want to read about cars and trucks, and not all girls want to read about dolls. These gender stereotypes are also being shattered through picture books that help children to understand that social norms are confining and that they can find topics that are appealing to them.

Find books that help your child explore new places around the world. Over 40 percent of children state that they want books that help to explore and understand new places they have never been before.[20] You don't need a passport to read books about South America or Asia. Is your child interested in animals? Expand your child's knowledge of snakes (for example) that may be indigenous to North America versus snakes that exist in Africa or Asia. I really love the Who Would Win? series about animal strengths and weaknesses. These are for early elementary readers and are great informational texts to expand your child's interests.

Start with one or two books in a series. Books that are part of a series are very appealing to young readers because of the interest in following characters and becoming invested in their adventures. Many early readers now help build the reading bridge from pre-school into elementary school, with characters that young children connect with and love.

PART II

THE SIX STAGES OF READING ALOUD

FROM BIRTH THROUGH SCHOOL AGE

IF ANY OF YOU reading this book have been to the pediatrician's office with your baby, you are probably familiar with the Ages and Stages Questionnaire.[1] This several-page document asks parents to respond to questions about their baby's physical, social, and cognitive development. *Does your baby roll a ball? Can your baby sit up? Does your baby smile?*

I was excited about this questionnaire as a first-time parent. I could finally document how amazing my sweet, perfect baby was at home, and now the doctor would know too! With each stage, at each visit, I filled out the form.

Until it came to communication at one year old. I couldn't check any of the boxes for Jagan. I even considered lying. But what good would that do? And whom was I lying to anyway?

The questionnaire told me what my oldest child *should* be doing at 12 months. The nurse insisted it wasn't a test, just a way to understand and help guide Jagan's development. But he wasn't saying even three words. He didn't say the typical "dada" and "mama," even though he would point to what he wanted, engage with us, play games, and listen attentively when we read aloud. The nurse said this wasn't a cause for concern, but there was a speech delay and it might help if I read books aloud to him.

I share my parental experiences with Jagan here (again) because we read multiple books to him every day. How can there be a correlation to between how much we read to him and having a speech delay? What does that even mean? How do I as a parent know when one stage starts and stops, and another begins? And who cares if he doesn't say "goo goo, ga ga" at six months old?

My experiences with Madhavi and Arjun were vastly different. Madhavi was reading books aloud at four years old, and Arjun benefitted from being the youngest child by having both siblings read aloud to him since birth.

I have shared the vast differences among my three children and how often they were read to, and how their speech and language developed so differently, only to acknowledge that every child is unique, just as we are as parents.

These stages are meant to do the same. Remember that a stage of development is an "age period when certain needs, behaviors, experiences and capabilities are common and different."[2] They are guidelines for what to potentially expect, how to support your baby, and what to be aware of that may be of concern.

It is not my place to make any diagnosis or state that these are hard-and-fast outlines. They are *stages* of reading aloud, and stages sometimes move faster and sometimes slower. You may find that your baby responds differently at two months than at six months when you read aloud. As a parent or caregiver, sibling, friend, or relative, it is your power to adapt to how, when, and what you will read aloud. You already know the why.

The six stages of reading aloud outline the characteristics of how your child listens and responds to being read to, how he will continue to develop and grow his interest and engagement in reading, and how you can build on these interactions in a meaningful way.[3]

We know that the greatest time of brain development is between birth and age three, but learning doesn't stop when you turn three. We continue to draw from our environments to grow and learn about the world around us. So here I want to share more about the stages of reading aloud that were an inspiration for this book, but I also want to acknowledge that reading aloud doesn't stop once a child enters preschool or kindergarten. Reading aloud begins at home and continues through schooling and adulthood. You're never too old to read aloud, or be read to.

I admit that I am fortunate to have a houseful of books and children who can read to each other. Truth be told, my daughter (when she was four) would read to my youngest (when he was two) every night. By the time he got to kindergarten, he had the vocabulary of a second-grader.

The six stages of reading aloud are developed from the research of many neuroscientists, educators, researchers, and parents who examined the ways we should read aloud to our children. Though each stage is centered around a period of development, much has changed in how we interact with our children and how we read and read aloud. The ways we read aloud to a three-year-old today, often with an electronic device in hand, would have been incomprehensible in the 1990s.

In the following sections, I outline and describe the read aloud stages. I hope that you recognize that these are fluid and ever-changing, that reading aloud should continue past preschool, and that these suggestions are starting points to building that lifelong read aloud habit in your home.

4

THE LISTENER

_____

PREGNANCY TO 2 MONTHS

WHEN I WAS PREGNANT, I would always talk to my first baby, Jagan. I created nicknames ("my little frog"), sang ("Twinkle, Twinkle, Little Star"), and read aloud. Based on who was reading or talking, Jagan would move differently, kick or turn, or be completely still. But sometimes I wondered if he could really hear me at all.

As I shared in the introduction, your unborn baby can hear you and anyone around you. And once you bring your baby home, your baby is stimulated with new sounds that are a stark difference from the protection in utero. Still, your baby can recognize your voice and the voices of people who were talking around you during pregnancy.

In the first two months of life, just like during pregnancy, your baby is *always* listening.

Brainy Baby: Your baby already recognizes your voice more than any other sound. As you continue to talk to your baby, her brain is growing and her understanding of language continues to expand. Don't be afraid to ask your baby questions (both during pregnancy and at this stage). The change in your inflection and tone is heard. For example, "Are you tired, sweet pea?" ends with an upward inflection in your voice. Your baby hears this, and it is the first step in helping her to build language skills. You expand those skills during read alouds. She may not reply vocally, but watch how your baby reacts physically to your loving questions.

Key Characteristics of Your Baby's Development

Listening
- Startled by unexpected noises
- Turns head toward caregiver or familiar voices
- Recognizes mother's voice (or the primary adult speaking to the baby during pregnancy)
- Recalls and remembers nursery rhymes or stories told and sung before birth
- Identifies rhythms and tones of mother's voice

Verbal
- Cries—that's all; your baby will cry to tell you what he needs

Visual
- Has limited vision and poor color vision
- Prefers to see a human face, especially Mom (even with no makeup on)
- Can distinguish black-and-white imagery
- Cannot see small details

Motor
- Imitates movements of mother's mouth
- Sleeps a lot (thankfully!)
- Demonstrates mostly reflexive movements (such as sucking and grasping)
- Coordinates body movements with adult speech
- Turns toward or away from sounds
- Prefers to be held and swaddled

Types of Books to Read

- Books with nursery rhymes
- Books read before birth
- Books you can sing
- Your favorite children's books
- Board books with black-and-white patterns

The number one reason to read to your baby soon after birth is to make her feel comfortable and loved with your undivided attention. This special time of connecting and bonding will help to reinforce a safe space and environment and begin to build the experiences you will share reading aloud as your baby gets older.

The second reason is to begin a routine that will become a wonderful habit for both you and baby that will last for years to come. The sooner you start reading aloud every evening, the more likely it will become a necessary part of your daily routine.

The third reason is to get other adults and older family members involved in building bonds and creating a literature-rich environment. Though I may reference one primary caregiver who may encourage other readers to be included in the routine, let's be clear: family dynamics are not just Mom and Dad. Even if Mom and Dad are the primary caregivers (such as in my family), the primary reader may not be the primary caregiver. My husband was the one who read aloud to our children after giving them a bath. It became his special bonding time after I was with them all day (of course it was also my time to sneak away with a glass of wine and my favorite shows). The reality is that there are blended families, older siblings, grandparents, and even neighbors and family friends who are integral to our children's development and lives. Each of those important people will only strengthen your baby's brain, language, and social-emotional development with unique read aloud routines.

This first stage is called "The Listener" because the first building block of literacy is developing the ability to listen.

The ability to listen will develop and continue to be important throughout your child's preschool years, and will be the key to learning throughout life. As I write this, I realize how many times I have actually asked my children if they are listening to me on any given day. *Sigh.*

One of my favorite books to read with my children to help them just calm down, relax, and listen (and an all-around classic) is Mem Fox's *Time for Bed*. With a newborn, there are several ways you can read this book. You can read it straight through in a regular voice while your baby is either awake or asleep. You can also start at the first page with the intention of trying out your parentese through the voices of the animals. And the animals are so much fun.

After reading, "It's time for bed, little mouse, little mouse, darkness is falling all over the house," you can interject something like "Can you hear the little baby mouse go squeak, squeak, squeak?" in a squeaky voice. Here you are already starting questioning, and you're using an expressive voice to get and maintain your baby's attention.

You can ask similar questions and make animal noises all the way through the book. In this way you are dialoguing with your baby and giving him the opportunity to hear many, many more words than just those provided by the book. This book also has a lovely, lilting, repeating pattern in the text that makes it an ideal bedtime story.

When to Read to a Listener

Anytime. Your baby could be lying next to you, just waking up, or drifting off to sleep. It is up to you where and when you want to build the read aloud routine. You can even read while your baby is asleep because your baby can still hear your voice when sleeping.

At birth, although it may appear that your baby is asleep, she can hear and be stimulated by sounds. I know a lot of moms who loved to read aloud to their baby while nursing. It might seem strange, but when your baby is tucked into your arm, reading a book, or even reciting lines from a favorite memorized story, will create a lovely, calming environment.

A six-week-old baby can already have a read aloud routine. For example, after I nursed Jagan, my husband would change his diaper, swaddle him, and hold him while he read aloud. There was also conversation. He would say, "This is for you, little frog" (remember our nickname for Jagan), and continue with "Do you want to read *Guess How Much I Love You* by Sam McBratney?" (That was my husband's favorite book to read aloud.) I had similar read aloud routines with Madhavi and Arjun. While Jagan was playing with his trains, I would hold Madhavi in my arm and read aloud *Time for Bed* by Mem Fox. Sometimes I would even read aloud a book I was reading for my own pleasure. As she grew older, she would turn her head to me when I paused. It was another reminder that even if we think they aren't, our babies are always listening to us.

How a Listener Responds to Read Alouds

After a few weeks you'll begin to notice that your baby responds with body movements when you talk and read to him. When parents talk directly to their infant, he can move his legs and arms in synchrony with their speech. Isn't that amazing? Your baby will cry but shows you he understands by his movements.

What responses can you observe in a newborn? Depending on your infant baby's waking or sleeping state, you may see that your baby appears alert or that she moves in synchrony with your speech. Responses may be very subtle at first. But know that every word you speak is being registered in your baby's active, growing brain. The more you are tuned in to your baby, the more you'll notice these small, sometimes subtle gestures of acknowledgment.

When you read to your baby, observe his arms and legs as you read. At this stage, reading Mother Goose rhymes or other poems with rhythm, alliteration, and rhyming words will encourage your baby to respond physically. The loving tones in your voice will help even an eight-week-old baby to focus on you, even if just for a few moments.[1] The book ¡Pio Peep! (by Alma Flor Ada, Alice Schertle, and Isabel Campoy) has been recommended for years as a collection of lovely Spanish nursery rhymes and is often included in preschool classrooms.

Why read a collection of Spanish nursery rhymes if you don't speak Spanish? It's just an example of how to build language skills early on and begin to see how reading aloud, language, and music are all connected.

These traditional Spanish nursery rhymes are among the most popular in Spanish-speaking countries. Spanish-speaking parents and grandparents will remember these rhymes from their childhood. Alma Flor Ada, a renowned author of children's books in both English and Spanish, selected them. Most of the Spanish rhymes have established, familiar, chant-like rhythms. Babies will soon grow to recognize and love these familiar sounds. Each Spanish rhyme includes an English translation that works well for parents using parentese in English. If you read the well-translated and re-created rhymes in English, you can even begin to make up your own rhythms.

Book Recommendations

Guess How Much I Love You by Sam McBratney
Big Red Barn by Margaret Wise Brown
Pat the Bunny by Dorothy Kunhardt

These three books were staples in our household. As I stated earlier, my husband would read *Guess How Much I Love You*, starting with Jagan and then with each child. Though I always loved to read *Goodnight Moon* (and, like many parents, soon found myself able to recite the rhyming text by memory), I love the setting and playfulness of Margaret Wise Brown's book *Big Red Barn*. It's a great book to start reading aloud, with calming, rhyming text that introduces children to the animals on a farm. You'll find that it is also a standard in most preschool classrooms. *Pat the Bunny* was one of the first books that my mom bought for Jagan, and it was also a great kinesthetic experience to read aloud and have Jagan rub his little hand against the soft bunny's fur on the cover. Needless to say, this became such a favorite among all three kiddos that we ended up with multiple copies.

Tips and Techniques

You may want to read the rhymes repetitively because repetition helps your baby to learn the rhythm and sounds of her home language. If the home language is mainly Spanish, your baby will learn to recognize her parents' native language. The more words she hears in Spanish, the easier it will be for her to learn English.

¡Pio Peep! is not just for families that speak and understand Spanish. The lullabies are filled with charming sounds and inflections. Don't be afraid to sing or use different pitches. Trust me. Your baby is not judging your singing voice.

STEAM Activities: Touch

Massage your baby after a bath. Who doesn't love a good massage? The massage stimulates your baby's cognitive and physical development. And as you massage, talk about each body part to help build language development. For example, my daughter was born with her foot slightly turned because of her placement in the womb. Our pediatrician said to massage her foot every day to help it reshape, and to this day she loves to have her foot held. We would always ask, "Can I hold your foot?" and spoke to her. Soon she would stick her foot out to us and say, "Fooo." Even though your newborn isn't going to respond with the correct name for each body part, she is listening.

Create baby art that represents your favorite book to read aloud. Because my husband always read *Guess How Much I Love You* to our kids, we used that book. We made a little art project with each child that had both my husband's hand and the baby's hand printed in washable paint. We wrote the title of the book at the top of the art.

We still have those pieces of art today!

Reading Aloud Habits

Choose a time and place that is quiet and free of distracting noises.

Select whatever you want to read: rhyming books, your favorite parenting magazines, or the newspaper.

Make yourself comfortable with your baby in your arms on a rocker, chair, glider, or bed, or on the floor.

Try to read for a few minutes when your baby is alert and has been fed so he is calm and attentive. You can also read while you are nursing if you have sufficient support for yourself (I used to just lie sideways on the bed and sometimes even work on my laptop while nursing). You may also want to continue reading if your baby falls asleep. He is still listening.

If you have other children, invite them to be part of the experience. Read what your toddler likes while you hold your newborn. Older siblings do want to be part of any bonding experience they observe, and inviting them into a read aloud moment is a great way to show that this is a collaborative, welcoming experience.

When your newborn opens her eyes, pause and hold her close to you so she can look at your face as you read.

Find the right time to read for you and your new baby. Yes, you are exhausted, but changing position, burping, or waiting a few minutes can alleviate discomfort, and then you can resume reading.

As long as the bed is protected with large pillows and your baby isn't turning or rolling, laying your baby next to you on the bed is a great space to have him listen as you read aloud before bedtime. In this situation you don't have to hold the book and your baby, but he is safe and being read to.

Activity: Start a Read Aloud Journal

I used to love keeping journals. I encourage all my kids to write their thoughts, document their experiences, and keep a journal. Sure, it seems like an easy suggestion to "keep a journal," but imagine spending five minutes documenting how your baby reacts to you reading, including specific movements and even books she responds to. In the first month, your baby may just be calm, listening (of course) and taking in the experience of being read to. The second month, you'll notice her alertness and her visual attraction to bolder picture designs. By the third month, she is so accustomed to the reading routine that she expects to be read to and wiggles her arms and legs in anticipation as you read with expression.

Adding photos to your read aloud journal, including the names of books you read, will make a wonderful gift for your child when he grows up, or something he can share in kindergarten and first grade. This will be visual proof of your gift of literacy to your child. The real proof will be your child's reading and writing ability, extensive vocabulary, and love of learning.

The Listener is our first stage and a prequel stage in the process of reading aloud. The auditory system is fairly well developed by the sixth month of pregnancy. Your baby can hear your voice, as well as music and other sounds, in the womb. Studies show that after birth, babies recognize specific books, rhymes, or music heard in the womb before birth, and these familiar sounds are emotionally comforting to them. When you are reading aloud to your baby while pregnant, your baby is hearing everything you say.

5

THE OBSERVER

---

2 TO 4 MONTHS

READING TO A two- to four-month-old is a lot of fun. Every time you read aloud, you are nourishing your baby's brain with the words she needs to build a solid language foundation. Every word you read and speak to your baby is a gift of language. In addition, when you read to your baby at this age, you'll feel a special bonding through books that delight both you and your baby. The intimacy of reading aloud is something your baby will expect every day, right along with nursing, sleeping, and bathing. And the best part is that your baby will start responding physically and verbally.

During the second and third months, your baby's world begins to expand with new sights, sounds, and textures. Feeling comfortable and secure in his surroundings, he begins to smile and become more social. This stage is called "The Observer" because he is now observing as intently as he listens. When my niece was two months old, she would focus intently on anything in front of her, sometimes just a plain white wall (I admit, sometimes it was a little creepy to imagine she could see something we couldn't), but the fact that she could hold her gaze was incredible to watch.

You'll notice that your baby can lock her gaze into yours for many minutes at a time. This connection between parent and baby is crucial for a baby's development and pure delight for Mom, Dad, and grandparents. It was really precious to see my older two engage with my youngest in this way. When they would build train tracks all around the house, I would hold my youngest up in my lap to watch his older siblings, or they would peer into his bassinet, connecting with him through his observations.

 Brainy Baby: At this stage babies can perceive bold colors and see details more clearly. Their binocular vision and depth perception are quickly improving. The illustrations that baby can now perceive in board books will stimulate and help improve visual skills. When you read to your baby at this young age, you are promoting both language and visual skill development.

Key Characteristics of Your Baby's Development

Listening
- Pays attention to your voice's changing pitch, rhythm, and volume
- Loves to be talked to about whatever you are doing: changing a diaper, cooking dinner, painting; this helps build language development

- Loves to listen to lullabies (read aloud or invented), calming music, and songs; musical albums from Putumayo Kids bring lullabies and soft tones from around the world

Verbal
- Begins to take turns when talking or engaging with the parent; won't be full sentences (although I know some moms who insist their three-month-old can say "hello")
- Develops expressive laughter
- Coos, gurgles, and grunts (my niece was such a grunter at this age that my brother and sister-in-law called the pediatrician many times out of concern; the pediatrician finally told them to stop and that it was a completely typical part of this stage of development)

Visual
- Extends range of visual focus to more than 18 inches
- Follows movements of anything moving, such as a mobile over a crib (my oldest *loved* these)
- Is drawn to checkerboard patterns and bright, contrasting colors
- Begins developing binocular vision and depth perception
- Can see more details in books and illustrations

Motor
- Tries to reach out to objects that he sees but lacks the full coordination to grasp onto them
- Can move her head from side to side while tracking a person's movement
- Can lift his chin and chest up if he is placed his on stomach (my youngest did the opposite—when on his back, he would clasp both hands and lift his entire upper body up off the ground)
- Plays with hands and toes, sucks on fingers and fists

Brainy Baby: Your baby will love to engage in "dialogue" with you. As you talk to your baby (or read aloud memorized phrases in books you both love), hold her facing you. She will communicate with you through her eyes, hands, facial expressions, and tone of voice. Research has also shown that when you hold your baby and speak to her, then suddenly turn away, she becomes confused as to the sudden break in connection. Even though you may not realize it, your baby is watching you and waiting for your verbal and physical responses.

Types of Books to Read

- Books with rhymes and songs
- Bold color or black-and-white picture books
- Books that you have been reading since birth that create a special moment or connection

Reading aloud during this stage focuses on "dialoguing," or taking turns talking to each other. This was introduced in chapter 2. Every parent of babies around this age knows the joys of the back-and-forth verbal and nonverbal interaction. This back-and-forth interaction that sometimes includes questioning is the basis for future social conversational skills. The intimate dialoguing between you and your baby not only promotes attachment but also is part of the way your baby learns how to identify his home language in the first months. Babies can actually distinguish between the language of their parents and other languages very early on. Babies listen to volume, pitch, and rhythm at this stage, and it's these qualities that help them identify the language spoken at home.

If your baby is positioned so she can see your face and the book, here are some of the things you'll see her doing. Your baby is now so alert that she'll be looking back and forth between the book and your expressive face as you talk and read. Although she won't fully understand the meanings of the words you're saying, she will connect with your mood and emotional tone. You should enjoy making sounds and voices, changing your pitch, and even singing as you notice your baby's reaction to the way you dramatize a story. She may show her excitement and engagement by breathing faster, shaking her arms, and looking back and forth between your face and the bold designs of the book. Your baby's verbal responses to your reading and talking will be in the form of sweet little coos and grunts. She'll also attempt to imitate your talking by moving her mouth in as many ways as possible.

Remember, the second stage in reading aloud is called "The Observer" because your baby's visual development is expanding to match his aural development. He will now become more engaged in your facial expressions and how those expressions match the tone of your voice.

This is a read aloud habit that will become even more beneficial when your child enters school. As a second-grade teacher shared with me about how she uses expression in class: "One of the things reading aloud does is help children to understand expression and punctuation. So if I wrote, 'Help Fire,' and we modeled punctuation—if I put an exclamation mark or a question mark, how does that change? And then one kid said, 'HELP FIRE!' And I said, 'Well, that is an important feeling.' And then I put a period instead, or comma, and we read aloud the differences. I taught them how important it is to consider punctuation when reading aloud. I teach them how the punctuation part helps them understand the feeling."

When to Read to an Observer

You'll be amazed at the length of your two- to four-month-old's attention span during read aloud time, assuming she is alert and has been fed and changed. (Taking pauses to change a diaper isn't a bad thing—it expands attention span and reinforces the quality time you are spending with your baby.) During this stage it is best if you speak in whole sentences so your baby can absorb the flavor, rhythm, and tone of your language.

By this age your baby is probably settling into a routine with sleep and feeding. (I say this with a lot of grace because my three children all had different routines, no matter how hard I tried.) However, my husband and I both loved reading aloud to our children right before bed. With multiple children in the household, I would cradle my youngest, Arjun, in my arms and read aloud while my husband read to our older two. Sometimes my daughter, Madhavi, would come lie next to me too, helping to build a cozy read aloud routine.

How an Observer Responds to Read Alouds

At this age your baby may reach out and try to grab what you are reading. When this happens, it is actually a lot of fun to watch. This instinct to grasp and touch will eventually lead to your baby holding a soft book or starting to turn the pages with you in anticipation of what might come next.

At this stage your baby will be attracted to black-and-white illustrations or bright and colorful designs, so books with this kind of art are the best choices. In addition, your baby will enjoy the rhythm and emotion heard in your voice from books with rhymes or those that can be dramatized with laughter or hand motions, such as *Clap Hands* or *Counting Kisses*. Your baby is learning the first steps in the give and take of conversation that the books listed below help promote.

Many of these books lend themselves to hand motions and physical touch and play with your baby. The challenge is to find a position that allows your baby to be placed in a car seat or sling, or on a bed, to free your hands for the hand motions. Your baby should be close enough to see clearly your face, your hand movements, and the book illustrations. If you're just beginning to read to your baby for the first time at 2 to 4 months, I suggest that you add some or all of the books from stage 1 because they include many well-known nursery rhymes that will become your baby's favorites for years to come. He can now enjoy looking at the illustrations in some of these rhyming books.

Book Recommendations

Playtime, Maisy! by Lucy Cousins

Maisy is an adorable mouse who is featured in many board books by Lucy Cousins, such as *Maisy's Favorite Animals, Maisy Goes to School, Maisy's Bedtime, Maisy's ABC, Happy Birthday, Maisy*, and many more. Open this book to any page, and you'll see why your baby's eyes will be riveted on the simple, colorful illustrations. This particular Maisy book is wordless. That means parents can talk about the mouse, Maisy, and what she is doing. Cousins's artwork exudes a sense of joy, a quality we all want children exposed to. This book also comes in a fabric version. It's light and soft and can be put in your baby's mouth—it's also washable. Toward the end of this stage, your baby will be trying to handle the book and also teething. Sturdy fabric or board books are great for this reason.

Counting Kisses by Karen Katz

This is a story that could be read at any time of day because your baby's smiles and body movements will tell you she wants to hear and see the book again and again. With its counting cadence—"Ten little kisses on teeny tiny toes, nine laughing kisses on busy, wriggly feet," and so on—it reinforces both basic counting and naming specific body parts. This book is simply fun to read aloud, and don't forget to kiss your baby with each page turn!

Tips and Techniques

The bright illustrations and bold colors in *Playtime, Maisy!* are appealing to your baby's visual development. How do you read this book? You make up the story! This gives you an opportunity to engage with your baby about the activities in the book that you might try together and what you like best. You can even insert some fun vocals: "Oooh! Maisy goes down the slide!"

Don't just take my word for it. Here is an actual review of this book from a parent who bought it online:

> *My two-month old LOVES this book, as well as the other cloth Maisy books. It is the perfect size for him to hold, and since it is so light he can play with it on his lap without me having to hold it for him. When I show him this book, his eyes get wide and he bats at the pictures with delight.*[1]

I will admit that some of the harder board books have gently landed on my children's hands or tummies while I was trying to find a good position to read to them, which is why being in a comfortable position is so important. It is also why books come in softer materials for younger ages. A soft, cloth-based book that can be smushed, manipulated, and chewed on actually helps to create a calming (and stress-free) reading experience.

STEAM Activities: Hear and Touch

Helping your baby to feel, hear, and experience multiple sounds and textures will support his language development. The book *Shake Look Touch* from Scholastic is a wonderful, soft, squishy book that has bright colors, crinkly materials, and soft, fluffy balls that will engage your baby during reading. This book also reinforces repetition and rhyme, which are important parts of developing receptive literacy skills. What other things can you shake and touch around your home that are safe for your baby to play and interact with while you are reading aloud?

Books that count simple numbers are also great for supporting brain development and teaching early math skills. *Counting Kisses* by Karen Katz is a great way to build a connection with your baby as you kiss her with each turn of the page!

Reading Aloud Habits

Establishing a quiet time and place to read will help balance the chaos of being new parents (or parents of multiple children).

Find a place that is comfortable and safe. Your baby will be observing you, engaged with how you talk and the book you are holding.

Remember that your baby may reach out and try to grab the book. If the book is soft, let him hold the book and manipulate the pages as you read.

Many of us are on the go, and audio books can be a valuable way to continue the literacy-rich experience we have at home while we are driving.

Activity: Create a Read Aloud Baby Book

Many of us know that we love to take pictures and document each stage of our baby's life and development. This starts with the fancy photo ops that come to you while you're still in the hospital. This "don't miss a moment" mentality is wonderful for documenting and cherishing each stage of your child's development. Let's be honest. The pictures you took for your first baby are triple what you are doing for your third. And that's OK. It's part of the journey of being a parent. Instead of getting overwhelmed with capturing, posting, tweeting, and texting every second, create what I like to call the read aloud baby book.

The read aloud baby book isn't just for your baby's handprints, a piece of one of her first outfits, or even a small swatch of hair. This book has pictures of your child engaging with you, your partner, a sibling, a friend, or even a pet during your read aloud routines. A picture of your baby in your arms while you rest a book on your lap, or your baby lying in bed next to you as you read, will soon expand to pictures of your baby reaching for the pages or sitting up and flipping through the book herself.

As the observer, your baby is watching everything you do and will continue to do so. Here, you are compiling a book of how you read together in different places and spaces. You are documenting how valuable the read aloud routines were throughout his life.

6

THE LAUGHER

---

4 TO 8 MONTHS

I HAVE A SHORT video of Madhavi at five months old resting on my lap after I just finished reading aloud a book. We were getting ready for her first flight, and she was fine. I was stressed. So I started talking to her in a silly, high-pitched voice, asking if she was: Ready to pack! Ready to go! Ready to help Mama!

With each question/exclamation, she clasped her hands tightly, looked directly at me, and burst into laughter. And if I was too slow with the next question, she paused, waiting and anticipating this next interaction. And as she waited, she watched me, learning to understand my facial expressions, actions, and dialogue with her.

Books for this stage need to cover a lot of ground and need to be entertaining. The books you read aloud need to introduce basic vocabulary that is related to your baby's environment and experience. Hearing lots of words is important at this time, because your baby's brain is storing words in her memory bank. Books that discuss a typical set of baby routines, such as bath time, eating, and bedtime, will help her hear vocabulary labels for each activity. Even without the book, the dialogue Madhavi and I had reinforced her new stage of development in laughing, enjoying our moment of informal conversation, and replying to me.

At age 4 to 8 months, babies are also attracted to books with textures and flaps for sensory exploration. Now that your baby is beginning to grasp, he will appreciate exploring the sense of touch. A book with crinkles, mirrors, and flaps will awaken your baby's senses all the time. Reading aloud becomes an incredibly kinesthetic experience, and you'll be surprised at what your baby finds funny during a read aloud.

Key Characteristics of Your Baby's Development

Listening
- Recognizes own name
- Can distinguish between the happy, sad, or angry tones of a parent's voice
- Understands inflections in voice
- Can tell the difference between the language spoken at home and other languages
- Absorbs and memorizes large numbers of sounds and words that will form the foundation of later speech
- Understands rhythm and tone of home language
- Likes to hear about daily routines such as diaper changing, feeding, and bathing
- Recognizes a few words

Verbal
- Vocalizes when making eye contact with parents or other stimulating objects
- Makes babbling sounds that emulate spoken language

- Imitates the pitches in your voice, especially if you are singing
- Will begin to coo, laugh, and giggle (my *favorite* part of this stage)

Motor
- Likes to examine different textures and materials
- Puts everything in her mouth
- Reaches for objects, especially books; may even try to turn the pages
- Begins to sit up and perhaps falls over gently
- Can pick up small objects with thumb and forefinger
- May turn and roll across the floor
- Lies on stomach and "swims" in place

Visual
- Looks intently at people and objects in the environment
- Gains interest in lights being switched on and off
- Sees colors and details clearly

 Brainy Baby: At six or seven months, a baby begins to make sounds that resemble spoken words such as *mama, papa, dada*. Remember that not all babies say the same words first—it depends on how they are spoken to and what is being read aloud. My oldest son's first word was *Bharat* (my husband's name), my daughter's first word was *grapes* (her favorite fruit), and my youngest son's first word was *mama* (yes, I said it over and over again to make sure he said *mama* first). What was your baby's first word, and what speech and language experiences influenced that moment?

Types of Books to Read

- Books that stimulate touch and feel
- Lift-the-flap books with hidden images and characters

- Teething books (board books that your child can chew on work too)
- Books with big and bright illustrations
- Books with repetitive words and phrases
- Books that label objects, toys, and parts of the body
- Books containing words and pictures about daily routines, such as bathing, eating, and sleeping
- Large books with colorful pictures and sturdy pages
- Plastic or cloth books that can be used in the bathtub or put in mouths
- Books with illustrations well matched to the text
- Books with buttons to push that create animal sounds—a great stimulant for this age range

When to Read to a Laugher

Your baby is developing an exponential rate. During these four months, your baby will start teething, become restless, or gain more focus on a particular activity.

You may find that your baby is restless or not focused. The best way to read under challenging circumstances is to get your baby's attention by using your voice at a higher or lower pitch than usual and dramatizing the words in a way that makes you and your baby smile. For example, it's hard for babies not to stop everything and watch and listen while you call out sounds like "caw, caw," "hee-haw," or "cock-a-doodle-doo" from *Barnyard Banter* by Denise Fleming. You can extend the fun later when you are in the middle of feeding or bathing by suddenly making some of the animal noises from the book. Notice how your baby will look at you as he begins to recognize these familiar sounds. Just wait until he starts to imitate them too!

You might also find that your baby is more restless due to teething and the fact that she is more mobile. You may have to change reading positions frequently or even give your baby a book to teethe on (yes, this is OK and even recommended). If your baby reaches for the book you are reading aloud (and it is sturdy or soft and squishy), allow her to hold and manipulate it. Just because your baby is active or uncomfortable doesn't mean you should give up on your reading routine. Keep reading!

How a Laugher Responds to Read Alouds

One of the most joyful experiences with reading aloud to your baby at this stage is the laughter, giggles, and vocalizations that will become a part of your read aloud experiences. Books that include animal sounds are a great way to engage your baby in vocalizing.

You'll observe that your baby will reach for books that he wants to read and listen to you read with greater focus than in previous stages. Because your baby is now more alert, he will respond to your voice, tone, and pitch. If you make a funny animal sound or change the inflection in your voice, your baby will turn to you as a reaction and soon begin imitating those sounds as well.

Book Recommendations

Dear Zoo by Rod Campbell
Where's Spot? by Eric Hill

Dear Zoo and *Where's Spot?* are great read alouds for this age. The sturdy board books and easy-to-lift flaps will engage your baby as you read. Please note: my son used to lift some of the flaps so vigorously that soon they just ripped off. Needless to say, we had multiple copies of *Dear Zoo*—some had the flaps in their original state, while others had taped-on flaps or flaps that were left off completely. These are also great books to continue dialogic reading. Pause before you lift a flap and see what your baby does in anticipation.

Tips and Techniques

As you read more books and build your library, remember to describe each object in the illustrations. Some books will have more words than others, but this is a great way to continue building vocabulary while engaging in a dialogue based on the pictures. "Look at the balloons! Which one is your favorite?" you may ask your baby, and then wait for a response. In this one reading session, he will be hearing hundreds and hundreds of words, your informal dialogue, and your excitement about the story. Think of it as food for the brain.

Your baby is more and more tuned in to illustrations because of improving visual skills. Choose books that stimulate a range of senses. Mini board books are tiny little books that feature illustrated objects with one- or two-word labels. They measure about three inches square and easily fit into little hands. Some have exceptionally good quality, such as those published by DK. It's a delight to see your eight-month-old sitting on the floor holding a mini board book, turning the pages just like an older child.

By eight months, babies are beginning to know many words that they can't yet express. You can read the book as is, or you can point to common items and name them: chair, curtain, table, door, bathtub. Pointing and naming helps reinforce your baby's vocabulary development. Soon your baby will be able to point to the chair when you ask, "Where is the chair?" When you ask your baby to find what's under the flap, emphasize the concepts of under, behind, and in as you answer your own question, modeling the correct language.

STEAM Activities: Water

Bath time is a great time to read aloud. There are soft, water-resistant baby bath books that your baby can hold during bath time. Even if there is only one word on the page, such as "water" or "bubble," reading during bath time is another way to build connections to science and language.

Books that have sounds and varied textures are appealing to your baby at this stage. Jagan's favorite book for almost two years was *Find the Frog*, a book I found in an airport gift shop the first time I had to fly and leave him for work. We would read this together every day, and I would use his hand to turn the flaps. Soon he would hold the book himself (when he could sit up a little) and flip the pages. I would also bring little bowls of water to splash and play with while we read, even adding toy frogs to make the reading experience come alive.

Reading Aloud Habits

If you started reading aloud from birth (or during pregnancy), great! Then you probably have a solid and established read aloud routine. Many of you will find that reading aloud before bedtime is convenient and easy to integrate into busy schedules. Research has shown that most families read aloud together at bedtime, but that's obviously not the only place to read aloud.

Read aloud during meal time. As your baby learns to sit up and starts to teethe and explore new textures in her mouth, reading books about new foods is a great way to connect reading aloud to her daily routines.

Reading in the car (if you're not driving) is a nice way to bond with your baby and engage with him. Because your baby is already seated and safe, place the book in his hands as you read aloud, encouraging him to turn the pages as you dialogue with him about the book. We always had tons of baby books and picture books stored in the back seat pockets of our car.

Many audio books available for your baby to listen to while you drive also include music and sounds as the story is read aloud. This is a great way to expand the read aloud experience. I share more tips about utilizing technology in chapter 12.

Activity: Create a Baby-Book Bin

Bookshelves are appealing and an easy way to stack and collect books, but they can also be dangerous if they are too high, over-stacked, or not fastened to the wall. Instead create a baby-book bin in your baby's room. Place the bin on the rug or in a location that your baby can crawl to and reach the books. A cloth basket or something softer that your baby can tip over or reach into is ideal. Fill the bin with books that you like to read aloud every day, and add to the books as your baby gets older.

When my daughter was around eight months old, she loved to crawl to her soft bin of books, reach in, and pull one out, or some-times two or three at a time. I would engage with her by asking, "Baby girl, do you want to read about puppies?" (if that's what the book was about), and I would either lie next to her on the rug or pull her to my lap. Before long we would have read through two or three books and spent 30 minutes reading together. As she got older, she would sit up and flip through the books or pull herself up to peek into the bin (the bins got bigger too). Reading and laughing together became one of our favorite routines.

7

THE BABBLER

---

8 TO 12 MONTHS

LET'S ASSUME THAT you have been reading aloud since pregnancy or at least since your child's birth. The ways your baby has responded to you have grown to include small head turns, grunts, laughs, and the possible tug of a book.

But everything changes now.

Here's why. At this time the many benefits of all the reading aloud you've done since birth will become apparent. You'll see how much your baby has developed a love and understanding of books by the way she can sit by a pile of her favorite books and look at them independently with no help from a parent or older sibling. She is capable of sitting up or lying on her tummy, crawling, turning pages, and focusing for short periods of time on the books that you have made a familiar part of your daily reading habits. If you watch, you will find her smiling, laughing, and engaging with the illustrations.

What's really incredible is how your baby will be able to recognize and identify favorite books. If you have been reading *Barnyard Banter* and make animal sounds in a different context (perhaps during playtime), your baby will crawl to the book bin and pull out the book. Jagan used to reach for *Goodnight Moon* and then start kicking his legs. Arjun would immediately vocalize when he saw any book that I would sing to him. This familiarity with books helps to establish the read aloud habits at home and will continue through the school age years.

Try this with the books you have read over and over again, especially after your baby is 10 months of age. Repeat a phrase from one of his favorite books, make an animal sound, or simply ask where his favorite character is. Then cheer as he crawls to find the correct book. Try this with some of the books you have read over and over, and observe how receptive your baby has been to your daily reading.

Key Characteristics of Your Baby's Development

Listening
- Understands (but cannot yet say) an average of 50 words by around 12 months
- Is developing ability to remember language that is heard repetitively from books or routines with parents

Verbal
- Can say most speech sounds
- Is beginning to make words at 10 months but will continue to babble beyond first year
- Imitates animal sounds when recognizing certain animals:
  Mom: Look at the cow!
  Baby: Mooooooo (or *maaaawww* or *mmmmmm*; you may hear many variations as your baby tries to build a reciprocal relationship with you)

Motor

- Likes to examine different features and textures of objects with hands and mouth
- Dialogues by gesturing, pointing, and verbalizing
- Tries to turn pages and points to pictures as you read
- Sits up in place, turns, and rolls around
- May begin to crawl, pull up against a chair or sofa
- May begin walking by 12 months (remember that these are guidelines: my three children were all at completely different developmental places in this stage)

Visual

- Sees colors and details clearly
- Has fully developed color, detail, and depth perception
- Has increasing control of hand movements
- Developing object permanence (more on that below)

 **Reading Rocks:** Create a homemade book about your baby's first birthday! Take pictures, draw images, place your baby's hand or foot in paint, and make homemade art. When reading, place your baby's hand or foot back on the handmade print to finish the book, reminding her that this book is all about her.

Types of Books to Read

- Books that encourage toddlers to chime in and repeat a word or phrase
- Word books that label objects, toys, and parts of the body
- Books that explore space and time concepts, such as inside, outside, under, after, next, up, down, tall, and short
- Books that illustrate action words, such as *running, jumping, sliding, sleeping,* and *eating*

- Books with flaps and noise buttons (be careful with these in public places like airplanes and restaurants; there are only so many "oinks" and "chirps" that people appreciate before their smiles change to frowns. I speak from experience)
- Specialty books with different shapes, textures, and sizes, with mirrors or noise buttons

As you can see every day, your 8- to 12-month-old baby is changing rapidly. This stage is one of the fastest in terms of development; your baby's brain, language, and social-emotional skills are expanding. He is now becoming a responsive, active member of your family, observing your face, listening to your voice, and imitating your actions.

These are the few months just before talking begins. Your baby is building a reservoir of words that, when you first hear them, will be cause for great celebration. The repeated words from every book you read will go straight into this reservoir. Some of the books I loved to read when my children were this age were books that can be sung, such as Raffi's *Wheels on the Bus*, or any books that build on connecting familiar nursery rhymes with text, such as *There Was an Old Lady Who Swallowed a Fly* and *The Farmer in the Dell*. At this stage you should read books with more complicated illustrations that encourage talking, pointing, and looking for details.

Babies are beginning to show their unique interests, and parents should try to select books that reflect these interests, such as zoo animals, construction vehicles, ballerinas, or dinosaurs. Try not to feel limited or restricted based on what you think your baby should be reading but instead offer a variety of books. My oldest son loved any books on trains, whereas my youngest son loved all books about bears. My daughter never cared about any books with unicorns or princesses. She preferred books with real animals such as puppies and kittens. She also loved reading books about dinosaurs with her brothers.

When to Read to a Babbler

I admit that while I am advocating for reading aloud any time you find time, when my first child was in this stage of physical development, I was exhausted by bedtime. And thus bedtime became the best read aloud time. My husband would come home and I would say, "Look! Jagan has been waiting to be read to by Dada."

Worked *every* time.

As my husband built this unique bonding experience with our children, our children benefitted from being read to, and I got a quick break. It was always a win-win situation.

This also reminded me of an *I Love Lucy* episode in which Ricky says he wants to tell Little Ricky "Little Red Riding Hood" as a bedtime story. Of course, Lucy, Fred, and Ethel all listen in. While Ricky recites the story in Spanish, he acts it out with grand movements, gestures, and facial expressions as the baby laughs in delight. This was television, and the baby probably wasn't there at all, but I challenge any of you to read aloud in that way!

Plus, this age group is so much fun to read to. You're not worried if they are sleeping or hungry or need a diaper change, because those routines have become more established. You can move from survival mode to enrichment—finding new books, topics, characters, and stories to build your baby's library and vocabulary.

How a Babbler Responds to Read Alouds

As you read aloud, notice how your baby responds with more engagement, dialogue, and physical response. Your baby will babble and start to "talk" back to you (in a good way) by emulating and repeating some of the words you say. Books with animal sounds or vehicle noises or books that you can vocalize are wonderful read alouds for this age group.

Research has shown that dialoguing is not just a natural part of how your baby begins to communicate with you; it is also the way your baby begins to build reading comprehension skills and practices replying and responding to your open-ended questions, such as "What did the Big Bad Wolf do?" Recall the different levels of dialoguing. Your baby may blow or make a raspberry sound, imitating how you "blow" when you read aloud. As your baby gets older, she will be able to expand her response to the same question, perhaps even with "huff and puff and . . ."

Book Recommendations

Chicka Chicka Boom Boom by Bill Martin Jr. and John Archambault

Why do I recommend an alphabet book for a child who isn't even one year old yet? Because I don't think of this fast-paced, rollicking rhyme as an alphabet book until preschool age. For an infant the letters are just a blur. However, your child will enjoy the rhymes and Lois Ehlert's bold, primary-colored graphics from this stage all the way through first grade. You will soon find the repeating verse of this book resonating in your head throughout the day: "Chicka chicka boom boom! Will there be enough room?"

At this age your baby loves to hear the same words again and again. By the time your baby reaches the age of two, he will know much of the rhyme by heart and may even be interested in some of the letters. It's also a really great book to read or recite in the car because of the rhythm of the text.

I always loved to read this book while any of my babies were resting on the floor on a blanket next to me. Showing the illustrations in this book is so important! You can also position your baby upright in a portable baby car seat or bouncer. You can hold the book so it's visible to you and your baby, and you can easily flip pages.

Through the repetition and rhyme, your read aloud is also building visual literacy. It is why this book is a staple in preschool classrooms across the country. As you read aloud, be as expressive as you can. Once your baby becomes older, you might even pause before the last word in the phrase "Will there be enough . . ." and wait for your baby to verbalize "oooo."

Good Night, Gorilla by Peggy Rathmann

If I had to list my top 10 favorite books, *Good Night, Gorilla* would be at the top of the list. I have seen a premature baby being read this book while in the NICU (neonatal intensive care unit) for two months. It continued to be one of this child's favorite books for the next two years. The story, which only uses the words *good night* and the names of the zoo animals, is mainly wordless. It shows the gorilla lifting the keys from the zookeeper's belt and unlocking all the cages. The animals follow the zookeeper into his bedroom. The story is just silly and fun! And minimal-word books such as this really allow for a lot of creativity and dialoguing between you and your baby.

Tips and Techniques

By eight months, babies can understand other words besides their own names. It's good for parents to remember that from this stage onward, babies will understand many more words than they will be able to say. Read books that are familiar and favorites, but don't be afraid to introduce books with more words and longer sentences. Books like *Corduroy* and *A Pocket for Corduroy* by Don Freeman still have engaging, bright pictures but begin to expand the vocabulary your child is starting to learn.

At this phase, nearly every day will bring a revelation of newly acquired skills, such as the ability to grab and hold objects, search for the objects, and even sit up (maybe even begin to pull up against the sofa) or hold the book while you read aloud. Find safe spaces for reading together, such as within a large play area or on the floor surrounded with cushions. Your baby will most likely not sit still when you are reading. She'll be finding things in her environment to engage with as you read, but trust me: she's listening to the story and loving it.

STEAM Activities: Textures

Collect soft materials and different textures from around the house. At this age your baby will love touching and feeling different textures. If you are reading a book about apples or different types of fruit, hold a piece and run your baby's hand over the textures. Help your baby to feel soft cotton and crinkly leaves as you read aloud. This creates a kinesthetic experience relating to the words he is acquiring.

One of the combined visual and physical characteristics of your baby now is that she will develop object permanence. This is one of the most exciting and engaging parts of a baby's natural development! If you are reading a book about balls, place a soft, squishy ball in front of your baby after reading, roll it to her, and then hide it nearby (under a pillow close to her). She will look for it! And when she does, reward her by saying, "Yeah! You found the ball just like the one in the book. Let's read the book again and look for the ball!" This activity engages the baby visually, physically, emotionally, and cognitively, and it motivates her to connect with you and the story.

Reading Aloud Habits

In addition to your baby's increased attention span, in the Babbler stage his vision is almost as mature as adult vision. Not only can he now enjoy the color and detail of the illustrations, but he can also understand a lot of words. You have probably noticed that when he hears words like *mommy*, *daddy*, *oma*, *opa*, *puppy*, *truck*, or *bird*, he looks in the direction of the person or object. He knows these and other words because you have repeated them daily when reading and talking.

Because your baby has been read to since birth, she will now sit on your lap and listen and look with concentrated interest at the book. She will want to turn the pages. She even knows how to wait to turn the page until you are finished reading or talking about it, or will push you ahead to keep reading if it is a familiar book. When you point at an object on the page, she looks where you are pointing.

As the name "The Babbler" implies, you'll notice that your baby can say most of the sounds common in your home language. You'll hear *da* and *gu* and *ba* and all kinds of variations. This is the same regardless of the language you are speaking at home. At this age (around 10 months), Jagan would put his hand out and say, "*tha-ma*" ("please give me" in Tamil), while asking to hold the book we were reading. Have you noticed any specific phrases or sounds your baby might be starting to repeat or emulate?

Your baby will also hold up his end of the conversation by gesturing, pointing, and babbling during your daily read alouds.

You, in turn, should listen and respond like you understand every word he says. He knows that you do too.

Parents are so tuned in to their baby's code that they often understand what her sounds mean even when others don't. I think this is the most remarkable aspect of parenting and watching babies develop at this age. How does a mother know that a certain sound or grunt her baby makes means she is hungry, or that her babbling is about a toy that she wants, when no one else would understand?

Babies have ways of letting you know their needs. Babbling of sounds is a crucial milestone on the literacy journey. The skill of babbling—and it truly is a skill—is the result of having heard your slowed, high-pitched, directed parentese speech since birth. As babies babble, they are learning how to distinguish individual sounds in words. This distinguishing of individual sounds in the stream of speech, called *phonemic awareness*, is tested at kindergarten and is a strong predictor of success in reading. Being able to hear and distinguish the smallest units of speech leads to letter-sound recognition and is an important part of reading and writing. Your babbler is on his way to becoming a successful reader, and he is just about to complete his first year of life.

Your baby can probably now sit up comfortably or even start to stand or lean against furniture. As your baby is exploring her physical environment, grab a book and read aloud. Make sure to do the following:

- Use a lot of drama! Your baby will love the changing inflections in your voice, the higher or lower pitches, and the physicality of shaking your head when you say "no" while smiling and laughing.
- Vocalize sounds. You've probably been doing this already when reading aloud: "What does the doggy say? *Woof! Woof!*" You will see now why this was so important. Your baby will vocalize back, imitating the sounds you make.

- Pause while you read. Remember that your baby's listening skills are expanding at an exponential rate. If you read a book with a predictive rhythm (such as *Chicka Chicka Boom Boom*), pause before you finish the final word or phrase. Even if you think your baby won't vocalize, you may be surprised by the reaction: the way he watches you in anticipation or responds with the actual word!

Activity: Make a Tear Book

At this stage babies love to tear and rip books. This might sound ridiculous, but I remember when I was pregnant with Jagan, I was watching an episode of *I Love Lucy* (yes, I do love Lucy) in which Lucy explained that her son (around 10 months old at the time, in the show) was "enjoying a good book . . . by tearing it." Sure, the audience laughed, but babies this age are moving from putting books in their mouth to exploring other possibilities (and limitations) of a good book.

Any time any one of my children tore a page out of a book, I encouraged it but also felt very guilty. I would desperately try to tape it back in or, if it was a book we really loved, purchase a gently used copy (in all honesty, we have four copies of *Where's Spot?*). Open-the-flap books are particularly exciting for young readers. They find the hidden object, but sometimes instead of closing the flap and putting the object back into hiding, they choose to rip the flap off completely. Can you blame them?

If you find that your baby is interested in turning the pages and ripping or lifting to find the hidden objects, make a "tear-apart" book with materials that can be torn and recycled.

Some parents might read this and think it is a ridiculous idea. Why make a book that your baby is going to tear apart? As an early childhood educator, I know that children during this stage of development have a heightened interest in exploring materials around them. This goes beyond books. Your baby can shred, rip, or tear anything! It's why, decades ago, we started babyproofing. The point is that it isn't bad to tear a book, but you have to consider what your baby is doing. It is part of his physical development to explore different materials (and his own strength). Isn't it better to have a set of materials that you made for your baby to rip than for him to rip the favorite book that you have been reading aloud together?

You aren't teaching your baby that it is OK to tear books; rather, you are allowing her to become comfortable in her own physical development and growth, and helping her understand that the books you two read together should not be torn.

8

THE BOSS

————

12 TO 36 MONTHS

CHILDREN BETWEEN THE ages of 12 and 36 months (one- to three-year-olds) are starting to formulate sentences, putting two or three words together in cute, funny, and sometimes meaning-less phrases that are left to us to decipher and translate. "Dat hoots!" could be *that hoots* (an owl?) or *that hurts.* Regardless, children are speaking or forming words at this point and, more important, they run the household. Literally. Run (and run around) the house.

Anyone who has a child knows this! It is a simultaneously challenging and exciting stage of development during which your child's brain is expanding every second. In these two years, your toddler's brain is growing faster than during any other period of development. It is also the time to continue to encourage an understanding of language, what yes and no mean, and what is safe, while engaging in a fun and meaningful dialogue of babbles, words, and laughter with your baby as you are probably chasing her around the house.

I also titled this section "The Boss" as a tribute to my lovely friend who said her daughter runs the house. And we agreed, Why not?

As long as she is running with a book.

Key Characteristics of Your Baby's Development

Listening

- Understands about 200 words; can say an average of 50 to 170 words
- Learns the structure of language, such as how to form a question and the proper word order of sentences (especially important is paying attention to inflection; even if the exact question isn't there with words, the inflection will indicate speech and language development)

Verbal

- Can say an average of 40 words at 16 months; understands 100 to 150 words (please remember that this is an average, and if your child is saying less that does not necessarily designate any issues with development)
- Imitates expressions such as "uh-oh" and "bye-bye"
- Begins to sing along to familiar songs and tunes (even by humming or creating his own language)
- Begins to combine nouns with verbs or prepositions to make two-word phrases or sentences: "For me?" (Jagan would say, "Tha-ma," which means "please give me" in Tamil)
- Begins to ask, "What's that?" and knows that objects have names; can name family members and point to them
- By 24 months, uses a word in different contexts; for example, will say "duck" when he sees his rubber duckie in the bathtub, a picture of a duck in a book, or a real duck in a pond
- May create his own language, such as saying "jebiya" for "jump" (this is the word Arjun would shout when he jumped up and down on the couch, bed, floor . . .)

Motor

- Responds to your questions with pointing, body language, sounds, and some words in an attempt to have a conversation
- Crawls, climbs, and walks; can crawl or walk to the bookcase and select favorite books (please be cautious of positioning books and bookshelves; as a teacher, I have heard about and seen too many accidents)

Visual

- Has improved visual memory, which is aided by the combination of rhymes or songs with movement, such as in "Itsy Bitsy Spider" (for example, when you see a spider and scream [or maybe that's just me], your toddler might start singing "Itsy Bitsy Spider" instead [that was my musical Madhavi])
- Can see and focus clearly on print, and may even begin to recognize specific letters (write out his name and hold his hand as you guide his fingers to trace the letters; soon he will point to the letters on his own, even if he is not speaking them)

 Brainy Baby: In this stage of development, your baby will want to do what she sees you doing: sweeping the floor, giving the dog a treat, or getting a book and reading it. I used to obsessively Swiffer and wipe everything, and when my youngest learned to crawl, he would always do so with a baby wipe in his hand. Think about what you are modeling with your behaviors and actions, and remember that you are your baby's first teacher.

Types of Books to Read

- Books that reflect your toddler's experiences, such as going to the park, visiting the zoo, putting on clothes

- Books that use phrases such as "goodbye" and "thank you" and that teach manners ("Yes, please!")
- Rhyme and song books that can be accompanied by hand movements, such as "Twinkle, Twinkle, Little Star" and "I'm a Little Teapot"
- Homemade books about routines using photos, drawings, or cutouts from catalogues or magazines
- Books about animals and their environments (farm, jungle, forest)
- Books about vehicles (trains, cars, planes, trucks)
- Books about items in the home (table, chairs, fridge)
- Books about potty training (and reinforcing to your baby, *Don't poop behind the curtain!*)
- Books about colors with corresponding words
- Books with one or two lines of rhythmic language on each page
- Books that label objects, toys, and parts of the body
- Books in different shapes, textures, and sizes, with mirrors or noise buttons (these are so much fun because they engage your toddler during the read aloud to press the corresponding button)

When to Read to a Boss

Someone on social media posted that putting a toddler to bed was like pulling the keys out of the hands of drunk person and chasing him around the room for three hours. Those of us with toddlers can relate.

This stage of development is so interesting. With my own three kiddos at this point, one was running across the room, one was still crawling, and one would stand and sit in place, terrified of falling or getting hurt. If I wasn't keeping one from flipping over a sofa cushion or planning dinner for us all, we would sit together and build, sing, and read.

Starting to establish a habit of reading aloud and reading together will become important at this stage. Many parents find that reading together in bed is the best and most convenient way to read. It's also a calming activity that will help you and your boss focus on a story and its pictures, as well as a bonding experience that will continue to develop and grow. And trust me, you will cherish these calm, focused moments with your baby.

How a Boss Responds to Read Alouds

Let's imagine this scenario. You are sitting in a cozy chair with your toddler in your lap. You have a great book that you are excited to read together. After a couple of pages, your toddler starts to squirm, stretches to get up, or wants to get down onto the floor and grab a toy. Maybe she is distracted. Does she want you to stop reading? No! Change the pace of how you are reading, or gasp. Then ask, "What's that?" Say, "Oh, no! Look what happened in the story!" She will turn and come to look. And even if she doesn't and is too distracted, you can either keep reading the story aloud, sit down next to her (wherever she has walked or crawled), or just leave the book within her reach.

Toddlers at this stage will take a book like this and go right to the page they are interested in. Notice what your child is interested in and either read the text or talk about what he is looking at. Or do both. Use the vocabulary in the book to ask questions during your conversations. For example, ask, "Can you find the forks in this forklift?" Your toddler will learn these labels quickly if this is a subject he is interested in.

Book Recommendations

Little Blue Truck by Alice Schertle

Toddlers love trucks!

You Are My Sunshine by Caroline Jayne Church

Try singing it too.

Little Owl's Night and *Little Owl's Day* by Divya Srinivasan

These are beautiful books about a baby owl exploring its world and getting ready for the night and the morning.

Chicka Chicka 1, 2, 3 by Bill Martin Jr. and Michael Sampson; illustrated by Lois Ehlert

This is a fun, rhyming companion to the popular *Chicka Chicka Boom Boom.*

The Snowy Day by Ezra Jack Keats

This is a favorite during cozy winter nights.

Tips and Techniques

Because your boss baby (wink, wink) is probably running the show, finding a quiet place and time to read together will become even more important. I have found that it can be difficult to get a toddler to focus and want to listen to me read. The amount of environmental stimulation can be overwhelming for a little one who can now crawl, walk, and probably run anywhere she wants.

Engaging your toddler with a toy or animal that is related to the book will help to make a connection to the story you are reading aloud. For example, having a fire truck, school bus, or toy train when reading a book about vehicles, or small dinosaurs when reading a book about dinosaurs, will help your toddler to focus on your read aloud and feel connected to the story.

STEAM Activities: Counting

Stack the Cats and *Balance the Birds* are two wonderful books that use clear, bright illustrations to teach children about basic mathematics. Why would introducing mathematical and counting concepts be important for a two-year-old? Because as you are building your child's vocabulary with every word he hears you speak and read aloud, your child is also learning to understand different math concepts that exist around him: size and shape, quantity and volume, and numbers.

If you think about it, some of the biggest and bestselling toys for this age group are always numbered or shaped to help teach basic math. Stack the blocks. Line the balls in numerical order. Even nursery rhymes (that we may or may not sing anymore) are filled with counting songs. A phenomenon that started as a You-Tube song, "Baby Shark," has led to multiple books that include counting and shapes and sizes. (I apologize for now putting that song in your head!)

Counting is a part of learning and preparing your child for pre-school. Books that include counting and basic math skills (including shapes and sizes) are an important part of building your baby's home library. As I have shared in other sections, having something physical (such as balls or blocks) that you can stack or balance as you read aloud will help to support a read aloud habit that keeps your toddler engaged and focused. The book *Goodnight, Numbers* by Danica McKellar is a wonderful bedtime book that includes basic counting skills through a lovely, playful rhyme.

Reading Aloud Habits

With three young children at home and my husband and me both working full time, I admittedly didn't always have time to sit and read—with them or even for my own pleasure! One of the habits I started was to leave books around the house, in every room. Just as a toddler will pick up a toy or musical instrument to play with, I found that my children began to pick up books that they wanted to read from these strategically placed piles. And, more important, I found that they would read together.

It also helped me learn what types of books and stories each child preferred.

Invite older children to read aloud to their younger siblings. I also found that this was a great way to ensure my older two children always felt welcomed to this experience. Soon they would read aloud to me and each other, and that became our read aloud habit.

Consider this conversation between an older and younger brother during a read aloud (my boys, of course):

J: Wook ("look"). The dinosaur is a T. rex, and the T. rex is big.

A: B—(*Babbles*)

J: The T. rex fights the *twiceratops*. They bahwal ("battle"). Roar. *Roar! Roaaar!*

A: (*Looks at J*) Ja-ja?

J: What you want to weed ("read") next, Arjun?

A: (*Points to another book, pushes one off the sofa*)

J: This dinosaur is a waptor. (*J is now silently reading. A looks at him, waiting, and then pulls the book from his hand.*)

What do you notice about the interaction?

Did the younger brother respond with words every single time? How did the older brother engage his younger brother? This interaction was between both of my boys when Jagan was four and a half and Arjun was 18 months. The piles of books (Who Would

Win series) were Jagan's favorites to read aloud to his brother. I have videos of how engaged Arjun was with Jagan's read alouds. Even if Jagan couldn't pronounce every word or read every word in the book, he would narrate stories based on the pictures, using dramatic sounds and actions to represent dinosaurs and animals.

And these two brothers bonded over books.

Activity: Create a Photo Book

At this age your baby is growing quickly! Like most parents and caregivers, you are probably taking a lot of pictures (most of which may never get printed). In our digital world, does anyone even print pictures anymore? I do! And I used these pictures to create a photo book of our monthly read aloud moments. I also expanded on this by creating our own counting book with photos of activities that we loved to do together.

For example, with Jagan, we loved to build trains together, lining up multiple coaches and tenders in different colors and patterns. Whoever made the longest train that could move around the wooden tracks without breaking was the winner! It was a fun game but also an expansion of what we would read aloud together, play together, and then remember together.

The photo book could also have images of your baby playing with her favorite toys. Adding words (handwritten or imprinted) creates a personalized book that expands your read alouds. At this age your baby loves to see images of herself, any siblings, and other babies. From a development perspective, your baby finds images of others like herself appealing and will try to interact or communicate with the images. It is also a way of helping your baby recognize and identify different body parts (eyes, mouth, nose, and so on), which is an important part of child development at this stage.

9

THE STORYTELLER

———

3 TO 5 YEARS

I MISS HAVING CHILDREN this age. Three-, four-, and five-year-olds *love* to talk. It's so different than my conversation with my now middle schooler (Jagan):

Me:	How was your day?
J:	Good.
Me:	Do you like school?
J:	Yeah.
Me:	Is anyone mean?
J:	No.
Me:	What did you eat for lunch?
J:	(*Big sigh*) Why do you *keep* asking me that?

I never had to ask, "How was your day?" any time after school when any of my three were in preschool. They would divulge any and every detail of the day. "Mom, did you know that we can go to the zoo to see zebras?" "Mom, Dani pushed a kid behind the shelf." "Mom, I was hungry but I didn't like the snack." "Mom, we read *Hungry Caterpillar* in class!"

Children at this age love talking and sharing about everything new that they are learning about in school, at home, or around them in the world. This is a time when you will see how reading aloud every day is influencing your child's development, understanding of language and text, and love of books.

And this stage is called just that, "The Storyteller," because your child will love expanding what he learns from books and creating stories about his own world. And if you're really listening, dialoguing, and maybe recording, you're helping to build a future writer.

Key Characteristics of Your Baby's Development

Listening
- Can focus on activities, read alouds, or other projects for approximately 30 minutes
- Is developing both expressive and receptive language
- Listens intently during read alouds
- May become distracted during longer read aloud times

Verbal
- Engages in dialogue and conversation
- Imitates patterns in speech and sound
- If learning another language (or in a bilingual household), can speak in both languages (or at least say phrases)
- Loves to sing spontaneously and invent songs
- Loves to read aloud
- Knows approximately 1,500 words
- Knows own name and is learning to spell first and last names
- Tells stories and retells events that happened during the day

Motor
- Is comfortable exploring physical aspects of development through running, jumping, climbing, and the like
- Will hold books, select books, and read books independently
- Can stand on one foot
- Uses scissors to cut simple shapes and patterns

Visual
- Loves to engage with illustrations in books
- Is drawn to picture books with images and topics that are a personal favorite
- Will make connections after reading through her own art or homemade projects

 **Reading Rocks:** This is the perfect stage to include books about numbers, colors, shapes, and vocabulary concepts that your baby will learn in daycare settings, during playdates with other friends, and soon in preschool and kindergarten. It is also very likely that these same books will be used in school, so you are already setting your baby up to be a strong reader in school during quiet reading times.

Types of Books to Read

- Books illustrating action words, such as children running, jumping, or sliding
- Books exploring space and time concepts, such as inside, under, after, and next
- Books that ask questions
- Books you can sing (my favorite is *What a Wonderful World*, illustrated by Ashley Bryan, which I also reference in chapter 10, about

how to connect books with music; even if you don't feel comfortable singing, play the song being sung by Louis Armstrong for a beautiful, colorful literacy experience)

- Books about your preschooler's current interests (get used to reading a lot about dinosaurs, trucks, trains, princesses, cats, dogs, and unicorns)

When to Read to a Storyteller

Your busy preschooler will most likely be attending some type of school-based or activity program, either through a local school, the park district, or a community center. Though there will probably be a group read aloud integrated into the school's curriculum, continuing to read aloud at home will help to build a home-school connection that supports and promotes reading aloud. Finding out what books the classroom teacher is reading is also important because your preschooler will come home and want to talk about the books read in class.

At this age bedtime is a great time to continue the read aloud habits that you have probably (hopefully) already started. The research shared in part I of this book has already shown that read alouds are typically done during bedtime, in a shared safe space with siblings, parents, or both. Find the book that your preschooler loved the most in school and make the home-school connection that is so important for your child's education. Show him that you have that book at home, that you two can read aloud together, and that you value reading aloud together.

How a Storyteller Responds to Read Alouds

One of the questions I often get from parents with preschoolers is whether it is OK to read the same book over and over again. While we run the risk of getting exhausted from reading the same story or memorizing the text until we can read it with our eyes closed, reading the same book every day is actually a good idea.

In the same way that babies and toddlers like to hear songs repeatedly, preschoolers also like and need to hear books read again and again. Repeated readings help children internalize the language. This is how their hearing becomes refined enough to detect rhymes and rhythms, which helps them acquire and later speak and read their own language. From repeated readings, your preschooler is gaining meaning of the spoken language, understanding text and context, and building a love of books.

I have talked to parents whose children have automatically taught themselves how to read at the early age of three or four. These children had so internalized the language of the books that were read to them repeatedly that they made the link to the actual letters and words on the pages in a way that allowed them to become early readers. Most children who are read to repeatedly as babies will learn to read effortlessly by first grade. And remember, even if they don't read fluently, they will be listening to you as you read aloud, helping to build and expand their vocabulary.

Book Recommendations

The Wonderful Things You Will Be by Emily Winfield Martin
Oh, the Places You'll Go! by Dr. Seuss

At this age your baby is now a preschooler and is learning about the world around her. One of the questions that is always asked in preschool classrooms is "What do you want to be when you grow up?" (To be fair, some of us adults are still figuring that out!) These books are classic picture books that engage children in playful rhymes and patterns, helping to inspire your preschooler to think about the possibilities that lie ahead.

These are also great books to engage your child in dramatic play and continue those conversations of "when you grow up . . ." that will begin in the classroom.

Tips and Techniques

Because your child is probably in preschool, daycare, or some type of social setting (even if only for a few hours a week), it's important to connect with the teachers to find out what books are being read in those settings. Having those same books in your home creates a strong foundation for building a love of learning as your child sees the connection between school and home through books.

You can also reach out to the teacher and share the favorite books that you and your child have been reading together. Volunteer to read aloud to the class (your child will be thrilled) and share the books and stories that have created beautiful memories at home.

STEAM Activities: Color

Children of this age love to engage with anything colorful. This includes painting, color recognition games, crayons, chalk, and even colored bubbles. Chalk and soluble crayons are a great way to connect language to color recognition. For example, using each colored crayon, write the name of the color and ask your baby to point to the word. Using the red chalk, write "red" on the sidewalk. Through this experience, your baby is connecting the color red with the word *red*, and you are reading aloud. Make sentences, write your baby's name, and have him trace the words too.

A Gift for Amma by Meera Sriram is a beautiful picture book that takes children through a journey of the colorful markets of South India. The colors on each page are vibrantly illustrated and connected to the flavors, spices, and sounds of an Indian market. After reading, connect to what sounds you hear, what scents you smell, and what colors you see during an afternoon at the park, in the backyard, or anywhere in your own home. Remember that every word counts.

Freight Train by Donald Crews is another wonderful book about colors. This beautifully designed, award-winning picture book begins with two or three large freight cars of different colors. The simple text (the color of which matches the color of the freight car) names the kind of car and color. Eventually, after each page turn, we see the entire train as it moves along and slowly moves past with a trail of smoke behind it. This was one of Jagan's favorite books to read together, which was fueled by his love of trains. With each page, he would find a train car from his collection and start building the train to match the order in the book. It was a way that we could reinforce his understanding of language and colors.

For a toddler, you will probably make train noises to accompany the text, or your toddler may do so for you. Imitating the train sounds with "choo choo" or "chug-a-chug" is a simple way to verbalize along with the text. It may seem silly, but trust me, the preschool teacher and kindergarten teacher (whose classrooms your toddler will soon be in) will do the same.

Reading Aloud Habits

One of the biggest differences in this stage of reading aloud is that your preschooler is probably in a classroom setting with peers who will influence her reading choices and teachers who will create read aloud experiences. If your child is being read to at school, do you need to keep reading aloud at home? Yes! This is the most important part of helping to build a lifelong love of reading.

Ask your child what books he read in school. What was his favorite book? What did the characters do? Many preschoolers love to read books about familiar and new characters. Books that take a simple, recognizable character like Pete the Cat are appealing to young children because they have simple story lines and bold illustrations and are a part of a series. You will find that the more children love specific characters, the more they will want to keep reading about them, look for their next books (if a series), and maybe even dress up and playact as them.

When my kids were in preschool, we used to love to read aloud together during dinnertime, bath time, or bedtime. We would take books in the stroller to the park or have them tucked in a snack bag or purse. Children at this age find it hard to learn to wait and have patience, and so wherever we traveled, we would have a few books to read together. This came in handy while we waited at restaurants, during takeoff or landing on an airplane, or even when stuck in the car in a traffic jam.

The best read aloud habit you can build for your preschoolers is to keep reading aloud, showing that books are important and valuable and that reading aloud together is something to love.

Activity: Create a Storytelling Coloring Book

Building on your preschooler's understanding and recognition of colors (and how to create and mix them), have her create her very own coloring book. This isn't any ordinary coloring book you might find in the grocery store aisle, with predrawn pictures. This is a book of colors that your child will create.

Stack four blank pieces of paper, fold them, and staple the folded edge to create a binding like a book's. At the top of each page, write just one word—each color—starting with *red*, then *orange*, then *yellow*, and so on until you fill up the book and each page is blank except for the name of the color that will be drawn.

After reading a book that highlights different colors, encourage your preschooler to use different materials to draw the corresponding color on each page. It doesn't have to be a red apple on the *red* page; it could be a scribble. You could even grab a red leaf and tape it to the page or, as my daughter did, use Mom's favorite red lipstick. Use simple sentences to create a story that accompanies each color. For example, "On Monday, we picked a red leaf." Now you are connecting with colors and using what is in your child's environment to create a personalized book.

I have to interject a side story here about colors: every time we went to Home Depot, my children were all fascinated with the paint-color wall, where they could each pick one or two swatches to take home. We didn't buy the paint, but the experience of seeing the rainbow of colors expanded their own vocabulary when we would read aloud. By the time my youngest, Arjun, was five, he would tell everyone his favorite color is cyan. I had to look it up.

PART III

STEAM READING

BRINGING ARTS AND TECHNOLOGY INTO YOUR READ ALOUD HABIT

THE OTHER DAY, I asked Jagan, "What does the word *STEAM* mean to you?" It was part quiz of what he knows and part quiz of what his middle school is teaching regarding STEAM. STEM/ STEAM curriculum is increasingly a central part of most public school education and an integral part of interdisciplinary learning.

Given his love of trains, I expected he would reference his large collection of steam trains. To my surprise, he spelled out the acronym: science, technology, engineering, arts, and math.

Me: Great! What is your favorite subject from all of those?

J: I love the arts . . . and math. (*Pause*) And engineering. And science and, oh yeah, I like technology too.

Me: (*Applause!*)

There's something amazing about the integration of all those fields of study and how they align with reading and literature. When you do a science project about building a volcano, you are being creative with the color of the lava and engineering the measurements, shape, and size of the volcano. For so long our educational systems have separated what is a natural part of learning and understanding the world. Though each of these subjects is separate and important, they are also interconnected, which is valuable for our children's learning and in helping them understand how to explore and interpret the world around them.

I want to highlight this important dynamic of how the subjects we learn in school are connected and also very often introduced to children through books. The series by Andrea Beaty that started with the bestseller *Rosie Revere, Engineer* is appealing to parents and teachers alike because it helps young children see that through trial and error and failure, you can still find success. My children's book, *Amazing Landmarks*, emphasizes the trial and error behind designing and building a structure as well as the teams involved, costs, and art choices. It helps young readers question the world around them and make connections.

That is what STEAM is all about.

Reflect on your own childhood. What is something you tried to build, design, mix, extract, draw, or create? Were you always successful? Did you get in trouble? Did you have to try and try again?

In the past decade, so much of what has changed in the world of children's books is ensuring that those stories are told. Just as we are making sure that children of color see themselves on the pages of books, publishers have also pushed to help young readers learn that making mistakes, trying and failing before succeeding, is an important part of growing. And read alouds are an important piece of that puzzle. How else can a four-year-old learn that it's OK if your art isn't perfect? Not everyone needs to draw the perfectly shaped pig.

Researchers have also examined how often parents and caregivers *actually* talk about reading aloud with other people. You would be surprised to find that only 25 percent of parents really talk about reading aloud and their reading aloud experiences with others, the challenges they face as parents, and just finding that balance between survival and enrichment.

To each of you reading this who has gotten this far in the book, I would *love* to hear from you. Share your stories, challenges, successes, fears, and delights. Remember that you are not alone and that your story about how you do or don't or simply want to read aloud will resonate with someone else.

In this section I expand on how STEAM is a part of your child's education and success in school, and how reading aloud and books at home can support his understanding of STEAM. I provide suggestions that relate to bridging music and art experiences through read alouds and expand on the topic of our technology-filled world by discussing devices that can read to your child (nothing to ever feel guilty about). Finally, in light of the COVID-19 pandemic, which has affected all parents and children around the world, I stress how we should always be prepared to support our children regardless of homeschooling, virtual learning, or open schools without masks.

I can't say that being isolated for months made me read aloud more with my children, but I recognized how much I missed taking my kids to the library, finding ways to expand their own sense of the world, and helping them to understand and value the shared experience of reading. What really saved us was the arts, the amazing advances in technology, and the time we spent connecting with each other.

10

USING MUSIC TO
SUPPORT READ ALOUDS

DID YOU KNOW that children sing before they can read?
Spend time observing your toddler or other young children in
a grocery store, elevator, playground, or any public space. Watch
how they turn a simple experience (swinging on a swing) into
a musical adventure. Observe how young children use music to
engage with and understand the world around them.

What is it about music and singing that engages young children? I could write an entire book on the importance of music
in young children's lives and the value of music and the arts in
supporting all aspects of development (and I have). I myself have
conducted hundreds of studies and published numerous books
on how music engages children and promotes academic learning
and social skills. And many parents and teachers know the value
of music without even reading one research paper.

A lesser-known fact is that music and singing can support
language development and phonological awareness (the ability to

recognize and manipulate sounds), which has been found to be a predictor of later academic success and reading ability.[1] This is different from recognizing sight words, a topic introduced in chapter 2. Phonological awareness is the ability to use the phonemes of language, the sounds made by letter combinations, to sound out a word. When a child cannot decode a word (or "sound it out"), it creates mental and emotional stress and can be a challenge that is hard to overcome.[2] This has very often been found in children who are diagnosed with dyslexia and have trouble reading and decoding words.[3]

Singing and using music can support children who have these reading delays and for children with speech delays or stutters. My younger brother struggled with speech and stuttering for years, and his therapist would use music, pitch, changes in tone, and singing as strategies for helping him. This is all part of building phonological awareness. One of the main parts of supporting children's phonological awareness is by helping them learn and distinguish rhyming words. For example:

> Cat, bat, hat, sat.
> Blue, shoe, shoo, you.
> How many more words can you add to each list above?

It's not just about how the word is written or spelled but also how it sounds and how it can be decoded. That's where music becomes so important. Think about how children are taught the alphabet. You sing! If you walk into any daycare or preschool setting, you will likely find the teacher or children or everyone singing: nursery rhymes, made-up chants about "cleaning up and putting things away," or good morning and goodbye songs.

This connection between music and literacy is a natural pathway for supporting read alouds with music. If you are going to sing a nursery rhyme, song, or folk song with your child, remember that there is probably a book or printed version that would make for a musical read aloud experience. Some of my favorite books to read aloud (and sing aloud) are *Baby Beluga* by Raffi and *What a Wonderful World* (I either played a recording of Louis Armstrong singing and we turned the pages together, or I sang the song aloud as my child turned the pages).

If you do an Internet search of "books that can be sung," you will find hundreds of titles. I love Jane Cabrera's *Row, Row, Row Your Boat* and *If You're Happy and You Know It!* When my children were younger, they also loved *The Wheels on the Tuk Tuk* by Kabir and Surishta Sehgal, which is a new take on "The Wheels on the Bus" focused on a popular form of transportation in India. Similarly, *Old Mikamba Had a Farm* by Caldecott Honor–winner Rachel Isadora takes young children on a journey through a farm in the African plains, playfully drawing upon the American nursery rhyme "Old MacDonald Had a Farm."

You may be self-conscious about singing or feel as though you don't have a good voice, or you may be a talented, gifted singer or even have a performance background. What I can guarantee after 20 years as a music educator and over 30 as a performer is that there will never be a young child who will question your singing ability. A classroom of preschoolers will be thrilled if you come in singing.

And your baby? She will be astonished, amazed, and engaged because you are singing a book aloud—and she will probably sing it back to you one day.

Begin by singing books that are already songs and see how that changes the read aloud experience. Your child is making connections between written, spoken, and sung words, hearing how words change and sound when spoken versus sung, and learning the different patterns in your spoken language. If you speak a language other than English, don't hesitate to translate the song even if the print is in English. This is an important part of supporting bilingualism and expanding how read alouds can help the development of a second or third language.

And when you get really comfortable singing and bringing music into read alouds, find your favorite book and make it into a song. It's like when my younger brother would always ask me, "Why do you have to sing everything?" My response was always, "First, you know I love music. And second, let's sing together." And we did.

11

USING ART TO SUPPORT READ ALOUDS

———

ILLUSTRATIONS ARE ALMOST as important as the text in baby books when it comes to generating language and inspiring the imagination. The words expose kids to rare words that you don't often use in regular conversation. The illustrations or photographs are helpful in stimulating your baby's visual development. Together these parts create literacy.

The best thing about the illustrations is that they encourage conversation. In fact, in some cases you don't even have to read the text, or there is no text—you "read" the pictures. Such is the case with Tana Hoban's *White on Black*, a wordless picture book, or the almost wordless *Carl's Afternoon in the Park* by Alexandra Day. After a few months of being read to since birth, babies can tell the difference when you're reading and when you're talking about the pictures. You may point to certain words for emphasis, like "no" in *No, David!* by David Shannon. Other times you'll point to the illustration of the tiny mouse in *Goodnight Moon* by

Margaret Wise Brown or the colorful keys in *Good Night, Gorilla* by Peggy Rathmann. Your baby will see you point to words and illustrations over and over and understand the different kinds of information you get from each.

When your baby is beginning to understand more, use Richard Scarry's *Best Word Book Ever*. Point to the illustrations and label them: *ball, shoes, pajamas*. This used to be Jagan's favorite book, even when he wasn't speaking the words. He would listen to me read aloud and point to items. He was soon able to point when I asked him to find each one, and eventually he would read the book to me.

There are a variety of illustrated books with or without labels. The wordless picture book *White on Black*, though simple, with one image on each page, is one of the first books babies see when they're introduced to labeling at two months.

Not that you expect two-month-old babies to know what you are talking about, but their vision and hearing are being stimulated as you comfort them with words and attention. You are introducing them to the process of reading pictures and having a conversation. Soon this process will be second nature, and they will know all about getting information from illustrations. At about 9 or 10 months, if you have read to your baby daily since birth, he will crawl over to the bookshelf or book box, take out his favorite book, independently look at the pictures, and turn the pages.

A great interactive activity to do with babies and toddlers after reading aloud is to create art that is inspired by the illustrations. The following books help to teach your baby about colors:

- *Mix It Up!* by Hervé Tullet
- *Mixed: A Colorful Story* by Arree Chung
- *Color Dance* by Ann Jonas
- *Planting a Rainbow* by Lois Ehlert
- *Five Colorful Crayons* by Lee Taylor

Using washable finger paint, create a memorable art project with your baby after reading. Sharing this project on social media and tagging the book title and author is also a great way to show how you are using art to expand the read aloud experience.

In addition to books such as *The Dot* and *Press Here*, *Tap the Magic Tree* is a lovely board book that encourages your baby to press each page and watch the tree grow. The text and imagery expand your baby's imaginative thinking, and if you take the book outside and read, have your baby actually tap a tree in the yard and engage her with observations about the environment (such as "look at the tree; hold the green leaf!"). You are expanding how your baby understands and explores the world.

You can start with an idea from any book to create amazing art activities. *Red Leaf, Yellow Leaf* and *Leaf Man* by Lois Ehlert are wonderful, classic picture books that will help your toddler connect with nature. Using an array of fallen leaves, you can make art with your baby. Here are some suggestions:

- **Leaf rubbing:** Place the leaf between two pieces of paper (any printer paper works). Rub an unwrapped crayon horizontally against the area of the leaf. If your baby is under 12 months of age, hold his hand and rub the crayon together against the paper. The image and texture of the leaf will appear on the paper as though by magic!

- **Leaf painting:** Create a painting station at home by protecting your table (and floor) with newspapers or tarp. We even have an old bedsheet that we put on the floor that dries easily for reuse. Place paint in small disposable cups (makes it easy for cleanup and for your toddler to dip and paint). Dip the leaf into one of the paint cups and use it as a paintbrush on plain or colored construction paper. As your toddler paints, ask, "What color? How does the leaf feel?" Connect the activity back to a book you have read together, either about leaves or colors.

- **Leaf collage:** Find an assortment of leaves in the park or your back-yard (remember not to pull them off trees). Toddlers love jumping in piles of leaves and throwing them everywhere. Arrange your collection on a piece of paper or canvas (you can purchase small canvases at a craft store or online for a reasonable price). Using a glue stick or drops of washable glue, help your toddler to arrange the leaves. Once the glue has dried, use a Sharpie to label each one by either color, shape, or size.

Art can also be a powerful tool for helping your child to express her understanding and interpretation of a book. A few years ago, I was invited by the Raffi Foundation for Child Honouring to develop a curriculum surrounding his song "Black Lives Matter to Me." As a champion of early childhood music and music literacy (books that can be sung), it was important for me to find ways to share the power of this movement in a way that was accessible and meaningful to the youngest children. Looking at how Raffi connected with families and babies through music, I coupled the message of his song and the power of the Black Lives Matter move-ment with art.

This isn't meant to be an advocacy speech or pushing an agenda. These are important events that your child will learn about in school, through social media, and through books. Your child's teacher will read aloud books that relate to important events in history, so finding ways to help your child understand and express his own understanding of these events is important.

And art is always a wonderful outlet for self-expression.

12

USING TECHNOLOGY TO SUPPORT READ ALOUDS

———

"**A**LEXA, TURN ON the lights."
"Alexa, open the garage."
"Alexa, read to me."

In the past decade, the ways in which we have used technology have changed dramatically. This is true of every decade, with changes from record players to Walkmans to MP3s to streaming music, and from phones in the home to in the car to in our hands. With the creation of smart devices and iPads, we can work, listen to music, watch shows, or read books wherever we go.

If technology is such an important part of our lives, why are we so hesitant to allow our children to use it? When we are on our laptops all day, why do we prevent our children from using devices? Is it really affecting their brain development in a negative way? Is it really that bad?

Growing up, my brother and I spent hours playing video games and watching television. Both of our parents worked full-time, and we often got off the school bus and let ourselves into the house. Who was there to dictate how often we could watch television?

In the past decade, many researchers have focused on how technology and screen time can hinder attention span and memory, noting that young children do not have the capacity to understand when to turn off devices and can become addicted to the shows they are watching.[1] But does that directly impact your baby's attention span and memory? Though some studies have shown that there could be a negative effect, this is not always consistently shown in the research findings. It is more important to consider guidelines rather than cut off technology and screen time completely.

For example, the American Academy of Pediatrics recommends that children below the age of two should not use screens at all (except when connecting with family on video calls) and suggests toddlers and preschool-age children (ages 2 to 5) should be limited to an hour of high-quality programming. The guidelines include the following:

- **18 months and younger:** Avoid screen time altogether for your child, except for calls with family.
- **18 to 24 months:** Choose high-quality programming to watch along with your child.
- **2 to 5 years:** Limit screen time to one hour per day of high-quality programming and try to watch along with your child.
- **6 to 12 years:** Place limits on what types of media are being used and ensure that they do not affect your child's daily behaviors or sleep patterns.

However, these recommendations are somewhat vague. What is high-quality programming, and who makes those determinations? In the past two decades, there have been YouTube channels dedicated to creating music videos, cartoons, and sing-alongs geared toward children from birth through age three. I admit that while some of them seem silly, they are appealing to this age group, with their bright colors and catchy songs. One of these songs, "Baby Shark," even turned into a global phenomenon.

In fact, some individuals make this their profession on YouTube, spending hours playing video games and giving tutorials to their young audience or creating videos in which they simply unbox toys. It seems absurd that a child would want to watch someone else open a toy and play with it, but in fact, all three of my children have found these videos exciting. (I have watched some myself, and I can at least say that these individuals have very nice manicures.) This becomes a marketing tool for companies, and those creating these videos gather lucrative endorsement deals and get items shipped to them for free.

Is it fair to say that a YouTuber has a short attention span? I imagine you need to have the ability to focus and concentrate if you're going to beat a video game in less than 24 hours, or spend hours editing for the correct lighting and angle for opening a Peppa Pig toy.

Based on my own experiences and childhood, I could argue that there is no relationship between screen time and attention span (or academic success) because my brother and I both have doctoral degrees and have never had any issue focusing in school (talking in school, yes, but not focusing, let's say, on a test).

Certainly many other factors influence attention span, memory, and academic success, but the irony is how the world changed during the COVID-19 pandemic. Preschoolers and kindergartners across the world who had never engaged with technology in schools were now sitting at home learning through a computer. My own children sometimes spent four hours or more watching their teacher lecture, viewing YouTube videos, or communicating with peers through Zoom, Webex, and other online platforms. What happened to all the advice on limiting screen time? And it makes me wonder, Did screen time become a necessity for education or convenience?

Even before the pandemic, parents around the world used technology to engage their children in educational apps or activities, to keep them occupied when grocery shopping or at a restaurant, or as a way to read aloud. Once we acknowledge that technology and screen time are a part of our daily lives, we can move toward what is meaningful, equitable, and appropriate. You don't need to limit television to weekends (if you do, that is your choice as a parent), but we should collectively recognize that technology (and screen time) is now integrated into our children's education.

Researchers have found that virtual read alouds, such as reading on a video call with Grandma, can be just as effective in supporting and engaging children's language and literacy skills as live book reading.[2] However, studies have also shown that virtual read alouds are not as beneficial to a toddler's language development, primarily because of how the parent interacts with the toddler.[3] Holding a book and reading aloud, pointing to pictures, and asking questions help to engage your toddler's brain, in contrast to playing a YouTube clip of someone reading a book aloud and turning the pages. Both are useful forms of read alouds, but it is important to consider what ways we can *effectively* utilize technology for reading aloud rather than moving away from new opportunities.

Besides being tucked in bed and read to by my mother every night, my earliest and fondest memories of being read to are through books on tape from the library. These cleverly packaged paperback picture books were cased in a plastic bag and accompanied by a cassette. Hung on racks across the children's section of the library, this was one of the earliest ways that technology was used to support and encourage read alouds.

If you are not familiar with these amazing forms of literacy engagement, you may find that a family member or friend has some at home. (One of my very good friends, who was a literacy specialist for decades, actually saved over 50 of these, with the metal rack.)

I loved this read aloud experience so much that I would use my dad's tape recorder and record myself reading a book. I even added the *ding* sound effect for turning pages. I would then force my brother to sit with me as we listened to my read aloud together. (I imagine he enjoyed this read aloud experience as much as I did.) It was also great practice for my future professional voice-over work.

The point here is that we were always using some form of emerging technology to support our children's learning (and even our own), whether or not we were chasing the latest exciting gadget. When nearly 65 percent of parents feel that technology is a distraction to their child's ability to focus or learn, while also acknowledging that our children are on devices anyway, why not use screen time for read aloud time?[4]

There are multiple innovative ways to effectively use technology to support read alouds and the read aloud experience. Here are some broad tips and techniques for using technology effectively for read alouds.

Alexa

It may seem odd or counterproductive to use an app or streaming service to read to your child. Ten years ago that might have been seen as bad parenting, but today your child is saying, "Alexa, when is the next Marvel movie coming out?" or "Alexa, how much does it cost to buy a horse?" (Maybe this is just my household here?)

In any case, the purpose of having a device or robot or read aloud through technology is that your child is likely engaged in technology throughout the day anyway. This is particularly true of school-age children who are now given iPads in school to use and take home, spend time on e-readers and reading apps in class, and are more technically savvy then we can imagine. It's like I tell my kids: if you're going to spend half an hour playing Minecraft or Roblox, you can surely spend half an hour reading, reading to each other, or in this case, having Alexa read to you.

Let's admit it. Young children enjoy having a device read to them. Perhaps by the time this book is published, robots will be able to scan and read a book aloud to your child as well. It's not as far-fetched as it seems.

So how do you utilize Alexa effectively? Geared primarily toward children ages 6 to 9 years old, the Reading Sidekick program engages children in shared read alouds. Your child can say, "Alexa, read with me," and select a book that he wants to read. This aspect of supporting your child's book choice is an important part of building lifelong readers and an innovative way that technology is being used to invite and engage children in reading.

We know that reading aloud and being read to stimulate your child's brain development, and when a child chooses the book she wants to read, she is more likely to continue reading and build a love of reading.

But is the experience of reading aloud with a device going to help your child understand what he is actually reading? One of the drawbacks of a product like Reading Sidekick is that there isn't any form of reading comprehension attached. Though the "voice" reading to and with your child might give praise and encouragement, engaging in a "shared" read aloud experience, there isn't the same level of dialoguing that happens when you are reading aloud—those informal moments of pause, reflection, and questioning that expand the read aloud experience. So until we get to a level where our devices can truly mimic the human read aloud experience, it is best to think of Alexa reading aloud as "edutainment" rather than education.[5]

E-readers

You know when you are in a grocery store and your toddler is in the shopping cart, and you just need her to chill for one minute while you pick that perfect cucumber? You hand her your phone.

You've done that. I've done that. The lady behind you is doing that. And that's OK.

Imagine, though, that you have an e-book pulled up that might read to your child, instead of your child watching a cartoon about babies with big eyes or videos of kids jumping in a ball pit. Even that five or ten minutes is a read aloud!

There are so many wonderful opportunities to build an e-library and add e-books to your smart device. Publishers are open and willing to encourage and support this way of reading aloud. It is a way of reading that isn't going to go away but only grow.

This is evinced in a video clip I watched recently that shows a teacher's surprise and delight when she took her first-graders to the school library with their iPads and they began to "mark" and list the books they wanted to check out and read. Expanding on that moment of choice, the children then began sharing their selections with their peers and encouraging them to also select the same books. At the end of the day, all of the children had almost identical reading lists on their iPads for e-books that they wanted to check out.

Hundreds and hundreds of books are now available to be read aloud as e-books or audio books. Some of my favorite children's books that have been translated into e-books include the Curious George books, which include music and monkey sounds and turn the page for your child. We have almost every Curious George book in paperback, but the read aloud experience through the e-books is something truly magical. The best thing about e-books is that you can now access hundreds of them anywhere, as long as your device is charged.

Apps for Reading

By the time my youngest was in kindergarten, I was already familiar with every type of reading app that school districts used across the country. Some were just to read aloud and engage children, and some were to measure reading comprehension (that is, your child would read the book and then answer a short quiz to see how well she understood the story). Many of these apps are only available for schools and librarians. If your child isn't in school yet, you can always try to gain remote access from your local library.

Some of my favorite apps are Raz-Kids and Sora. Raz-Kids is an app that is often purchased by a school district and then continually updated based on your child's reading level. Some teachers keep it open for children to read any level of books they want, while others limit the range based on what they believe the child should be reading. I don't agree with limiting children's access by reading level. Here's an example of why: Arjun's kindergarten teacher believed that all her students didn't need to read past level C and locked all the corresponding levels. The levels are alphabetized (starting with AA up to Z). Arjun was able to read chapter books by the middle of kindergarten, so limiting him to a book that had simple sentences such as *there is a cat* was not challenging him at all. He eventually got bored of the app (who can blame him?), and we stopped reading through Raz-Kids completely. The caution here is that you want your child to be challenged but not overwhelmed when reading aloud, to feel successful but not bored. It's a tricky balance, but apps like Raz-Kids can help you—and hopefully your child's teacher—to find that balance.

Sora is another app I was introduced to through my children's school. This app is connected to the local or school library and empowers children by allowing them to select the books they want to read, create a collection (by checking out physical books, for example), and then return the books when they are done.

Some of the research I have been working on with national organizations explores the power and ownership book choice gives young children. As the board chair for the nonprofit Bring Me a Book, I continue to explore the power of *choice*. What we know is that when given the choice to pick a book, a child is 50 percent more likely to read and keep reading.[6] And 89 percent of children said that their favorite books and series to read were the ones they selected themselves.[7] Remember that young children don't always have the opportunity or power to make choices in other areas of their lives, so by offering them the opportunity to pick what book to read (and when they enjoy it), we make them feel more and more empowered.

Similar to Sora, the app Epic! is available to schools and libraries and engages children in selecting books that they want to read or hear read aloud. Epic! also sends notices and e-mails asking parents to make recommendations for their kids, and those suggestions then pop up for students when they open the app in school. My kids all prefer Epic! to other reading apps.

Your local library may also offer hoopla, which is similar to Sora but allows you to check out any and all books in your library. I learned about e-library services from my mother, who is an avid reader and never without a book in her hand, but I found it fascinating how she continued to utilize the library even during the pandemic. While she couldn't go visit physically, apps such as hoopla allowed her to check out books she wanted and ones that were exciting and engaging for my children to listen to her read aloud.

If it isn't already obvious from these apps and suggestions, you can engage your child in read alouds without having to purchase a physical book. And that's OK. The ways we engage with technology have changed so much in the past decade, why should we not expect our reading, music, arts, and technology experiences to change as well? A great part of returning to the library or connecting with your local library again (if you haven't already done so) is that books you check out and read with your children through e-readers or apps will introduce them to new characters, stories, and series. And if you fall in love with a character or story (as we so often do), you can build that library in your child's room of both physical books to hold and those to read on a device. Ultimately, you can begin to build a library of books that you love to read aloud together (or that your child likes to have read to him) without ever leaving your home.

EPILOGUE

—————

RAISING A LITERACY-RICH CHILD

IN THE TIME since I was a child, the topics included in children's books have greatly changed. No longer are children's books focused only on children with blond hair and blue eyes but now include diverse families with multiple races or same-gender parents and children with thick, black hair or even a mustache. *Laxmi's Mooch* by Shelly Anand is a wonderful story about body acceptance, and *How to Wear a Sari* by Dharshana Khiani is a book I would have loved to read as a child.

In fact, publishers over the past decade have pushed to change the dynamics of what is put in print. The We Need Diverse Books movement coupled with larger activist organizations have also pushed for the inclusion of different images of children—those that represent our rich, diverse, and varied populations rather than those that mold or reinforce stereotypes.

Why is representation in children's books important? One major fact is that young children of color are not often represented in books, and certainly weren't when I was growing up. I wanted to be Sleeping Beauty when I was younger, but I didn't look like her (and still don't). Sleeping Beauty and the other princesses were only depicted with a light skin tone and blonde or brown hair, a homogenous representation of what was considered beautiful. As a result, I never saw myself as beautiful when I was a child, and I wonder: Would that have changed if Sleeping Beauty's family was from South India, like mine?

One of the ways children connect with the world around them and feel seen is through books. Representation is also important because research has shown that when children *choose* the books they want to read, they are much more likely to have a desire to read and invest in their own learning.[1] Encouraging self-selection also means that your child is reading books that match her personal interests, and having options that reflect what your child sees in the mirror allows her to do this. This also engages her in sharing reading experiences and books with her friends because reading is a social experience—whether your child connects with characters in the book, with you reading aloud, or with friends.[2]

Your child is more likely to pick up a book that her friend is reading in school or that a teacher reads aloud in class. Like one elementary teacher told me, after she finishes reading *Oliver Twist* or *The Miraculous Journey of Edward Tulane* to her second-graders, the book is always checked out at the school library. She also reiterated the importance of modeling, saying, "The thing about a read aloud and modeling, for those kids who are struggling with reading, even if they are not able to read, they are hearing the story, the words, the vocabulary. They can understand the inference being made. These kids can do cause and effect, predictions, everything we are doing as a group for second-grade readers, and even though they can't do it on their own, they are hearing it and it is being modeled."

We hope that good reading habits and reading experiences are included in our child's education in school, but never forget that you are your child's first teacher. He is watching everything you do and absorbing all your daily habits. And this part of teaching and modeling as a parent is perhaps the most important part of demonstrating the value and importance of reading.

My colleague Alan Boyko, who was the president of Scholastic Book Clubs for decades, coined the term "super reader"—in many ways a reference to himself as someone who reads every day and always has a book in his hand, and to those who read so constantly they sometimes don't remember what books they have read (that's my mom).

But it also invokes the idea that we *want* our children to become super readers. And everyone can become a super reader, regardless of race, education, and socioeconomic background. How you help to support your child in becoming a super reader is based on how you demonstrate the value of reading aloud at home and the ways you continue to build your home library of books that you and your child love to read together.

While a lot of brain-based research certainly fueled this book—ideas and strategies for supporting your child's overall brain growth, language acquisition, and social-emotional health through read alouds—I hope that there are takeaways that you can start to try out in your own home. So I want to provide a quick "top 10" list of how and where to begin:

Top 10 Read Aloud Tips

1. Build a library of books around your home.
2. Read books that represent you and your family to help your baby to be seen and feel acknowledged.
3. Have a separate bookshelf or book nook in your baby's room.
4. Expand books based on your baby's stages of development and interests.
5. Model what reading habits you want your child to have.
6. Encourage family to share read alouds through video calls or other forms of technology.

7. Include e-books and e-readers on your smart devices.
8. Follow each read aloud with a STEAM activity (a song, an art experiment, a cooking project, or the like).
9. As your baby gets older, encourage her to read aloud to you, pets, family, and friends.
10. Read aloud every day, even for just 10 minutes.

The research is clear: reading aloud supports brain health, and a stronger brain means a stronger mind and stronger body in old age. Do you wish you were read to more or wonder what you would be like if you had been? Is that something you would want to change?

You may even think that once your child reaches preschool age or is becoming more of an independent reader that you don't need to read aloud anymore.

Well, it's the exact opposite.

Literacy experts and elementary teachers found that it is even more important to read aloud as children become older, to continue modeling why reading aloud is important, share in new literary adventures, and help foster social-emotional development.[3] It's like I tell everyone who will listen: you are never too old to read aloud, or be read to.

Regardless of your own experiences, knowledge of children's books, or comfort with reading aloud, it is never too late to begin reading aloud to your child. Every book you read aloud supports your baby's brain development, language development, and social-emotional development.

To borrow a line from the character Dory in the Disney/Pixar movie *Finding Nemo*, "Just keep swimming, just keep swimming." I leave you with this one, broad takeaway: even when life gets crazy busy, even when you are overwhelmed, even if there is another pandemic and we cannot leave our homes . . .

hold your baby close, find a cozy space,
and just keep reading.

NOTES

Introduction

1 Claire Cain Miller, "Pandemic Parenting Was Already Relentless," *New York Times*, June 26, 2020, https://www.nytimes.com/2020/06/26 /upshot/virus-intensive-parenting-education.html.

2 Margaret K. Merga and Saiyidi Mat Roni, "Empowering Parents to Encourage Children to Read Beyond the Early Years," *Reading Teacher* 72, no. 2 (2018): 213.

3 Jim Trelease and Cyndi Giorgis, *Jim Trelease's Read Aloud Handbook*, 8th ed. (New York: Penguin Books, 2019).

4 Meghan Cox Gurdon, *The Enchanted Hour: The Miraculous Power of Reading Aloud in the Age of Distraction* (New York: HarperCollins Publishers, 2019).

5 "The Growing Brain," Zero to Three official website, accessed April 11, 2022, https://www.zerotothree.org/resources/1831-the-growing-brain -from-birth-to-5-years-old-a-training-curriculum-for-early-childhood -professionals.

6 "Early Brain Development and Health," CDC, March 5, 2020, https:// www.cdc.gov/ncbddd/childdevelopment/early-brain-development.html.

7 Betty Hart and Todd Risley, *Meaningful Differences in the Everyday Experience of Young American Children* (Baltimore: Brookes Publishing, 1996), 160, 198, 199.

8   Sharon E. Fox, Pat Levitt, and Charles A. Nelson III, "How the Timing and Quality of Early Experiences Influence the Development of Brain Architecture," *Child Development* 81 (2010): 28.

9   Lise Eliot, *What's Going On in There? How the Brain and Mind Develop in the First Five Years of Life* (New York: Bantam Books, 1999), 238.

10 Warwick B. Elley, "Vocabulary Acquisition from Listening to Stories," *Reading Research Quarterly* 1 (1989): 174; Terry Meier, "Why Can't She Remember That? The Importance of Storybook Reading in Multilingual, Multicultural Classrooms," *Reading Teacher* 57 (2003): 242; Elizabeth F. Pemberton and Ruth V. Watkins, "Language Facilitation Through Stories: Recasting and Modelling," *First Language* 7 (1987): 79; Claudia Robbins and Linnea C. Ehri, "Reading Storybooks to Kindergartners Helps Them Learn New Vocabulary Words," *Journal of Educational Psychology* 85 (1994): 55; Monique Sénéchal and Edward H. Cornell, "Vocabulary Acquisition Through Shared Reading Experiences," *Reading Research Quarterly* 28 (1993): 360; Grover J. Whitehurst, David S. Arnold, Jeffery N. Epstein, Andrea L. Angell, Meagan Smith, and Janet E. Fischel, "A Picture Book Reading Intervention in Day Care and Home for Children from Low-Income Families," *Developmental Psychology* 30 (1994): 679.

11 Ohio State University, "A 'Million Word Gap' for Children Who Aren't Read to at Home: That's How Many Fewer Words Some May Hear by Kindergarten," Science Daily, April 4, 2019, https://www.sciencedaily.com/releases/2019/04/190404074947.htm.

12 Scholastic, *Rise of Read-Aloud* (New York: Scholastic, 2019), https://www.scholastic.com/content/dam/KFRR/Downloads/KFRR_The%20Rise%20of%20Read%20Aloud.pdf.

13 "Read Aloud Survey Report," Read Aloud 15 Minutes, 2018, https://www.readaloud.org/surveyreport.html.

14 Adriana G. Bus, Marinus H. van IJzendoorn, and Anthony D. Pellegrini, "Joint Book Reading Makes for Success in Learning to Read: A Meta-Analysis on Intergenerational Transmission of Literacy," *Review of Educational Research* 65 (1995): 15.

15 Christopher J. Lonigan, Jason L. Anthony, Brenlee G. Bloomfield, Sarah M. Dyer, and Corine S. Samwel, "Effects of Two Shared-Reading Interventions on Emergent Literacy Skills of At-Risk Preschoolers," *Journal of Early Intervention* 22 (1999): 308.

16 Lesley Mandel Morrow, "Reading and Retelling Stories: Strategies for Emergent Readers," *Reading Teacher* 38 (1985): 875.

17 Diane August and Timothy Shanahan, *Developing Literacy in Second-Language Learners: Report of the National Literacy Panel on Language-Minority Children and Youth* (Mahwah, NJ: Center for Applied Linguistics, Lawrence Erlbaum Associates, 2006).

18 Scholastic, *Kids & Family Reading Report*, 7th ed. (New York: Scholastic, 2019), https://www.scholastic.com/readingreport/home.html.

19 Doris Luft Baker, Vivianne Mogna, Sandra Rodriguez, Dylan Farmer, and Paul Yovanoff, "Building the Oral Language of Young Hispanic Children Through Interactive Read Alouds and Vocabulary Games at Preschool and at Home," *Journal of International Special Needs Education* 19 (2016): 82.

20 Susan A. Everson-Rose, Carlos Mendes de Leon, Julia L. Bienias, Robert S. Wilson, and Denis A. Evans, "Early Life Conditions and Cognitive Functioning in Later Life," *American Journal of Epidemiology* 158 (2003): 1085.

21 Robert S. Wilson, Patricia A. Boyle, Lei Yu, Lisa L. Barnes, Julie A. Schneider, and David A Bennett, "Life-Span Cognitive Activity, Neuropathologic Burden, and Cognitive Aging," *Neurology* 81 (2013): 318.

22 Donna Schatt and Patrick Ryan, *Story Listening and Experience in Early Childhood* (New York: Palgrave Macmillan, 2021), 15–38.

23 Kumar B. Rajan, Rekha S. Rajan, Margaret Aker, Lisa L. Barnes, Robert S. Wilson, Neelum T. Aggarwal, Jennifer Weuve, Charles F. DeCarli, and Denis A. Evans, "Early-Life Experiences, Structural MRI, and Cognitive Function in a Population Sample" (in press).

24 Guilherme Brockington, Ana Paula Gomes Moreira, Maria Stephani Buso, Sérgio Gomes da Silva, Edgar Altszyler, Ronald Fischer, and Jorge Moll, "Storytelling Increases Oxytocin and Positive Emotions and Decreases Cortisol and Pain in Hospitalized Children," *Proceedings of the National Academy of Sciences of the United States of America* 118 (2021).

25 Susan Ledger and Margaret K. Merga, "Reading Aloud: Children's Attitudes Towards Being Read To at Home and School," *Australian Journal of Teacher Education* 43 (2018): 124.

26 Nicholas Zill and Jerry West, *Entering Kindergarten: A Portrait of American Children When They Begin School* (Washington, DC: US Department of Education, National Center for Education Statistics, 2001), https://nces.ed.gov/pubs2001/2001035.pdf.

27 Jeanne Bonner, "Many Turned to Libraries During the Pandemic for Free Wi-Fi and Other Services," CNN, June 14, 2021, https://www.cnn.com/2021/06/13/us/coronavirus-libraries-pandemic/index.html.

28 Valerie Nye and Christopher Schipper, "Long Live the Library," *Inside Higher Ed*, January 28, 2021, https://www.insidehighered.com/views/2021/01/28/libraries-have-been-crucial-during-pandemic-and-need-more-support-opinion.

29 "Impact of COVID-19 on Public Libraries," Hunt Institute, April 30, 2021, https://hunt-institute.org/resources/2021/04/impact-of-covid-19-on-public-libraries/.

30 "Impact of COVID-19," Hunt Institute.

Part I: Why Read Aloud? How Reading Aloud Supports Your Child's Development

1 Roberta Michnick Golinkoff and Kathy Hirsh-Pasek, *How Babies Talk* (New York: Penguin Books, 2000), 51–52.

2 "Pediatrics Group Wants Parents to Read to Their Children Every Day," *U.S. News & World Report*, June 24, 2014, https://health.usnews.com/health-news/articles/2014/06/24/pediatrics-group-wants-parents-to-read-to-their-children-every-day.

1. Benefits of Reading Aloud for Overall Brain Growth

1 Rajan et al., "Early-Life Experiences"; Rima Shore, *Rethinking the Brain: New Insights into Early Development* (New York: Families and Work Institute, 1997), 19.

2 Justin Worland, "This Is What Happens When You Read to a Child," *Time*, April 27, 2015, https://time.com/3836428/reading-to-children-brain/.

3 John S. Hutton, Kieran Phelan, Tzipi Horowitz-Kraus, Jonathan Dudley, Mekibib Altaye, Thomas DeWitt, and Scott K. Holland, "Story Time Turbocharger? Child Engagement During Shared Reading and Cerebellar Activation and Connectivity in Preschool-Age Children Listening to

Stories," *PLOS One* 12, no. 5 (2017), https://journals.plos.org/plosone
/article?id=10.1371/journal.pone.0177398.

4 Marian Diamond and Janet Hopson, *Magic Trees of the Mind* (New
   York: Penguin Putnam, 2000), 37.

5 Shore, *Rethinking the Brain,* 19.

6 "Pregnancy Week by Week," Mayo Clinic, June 30, 2020, https://www
   .mayoclinic.org/healthy-lifestyle/pregnancy-week-by-week/in-depth
   /fetal-development/art-20046151.

7 Trelease and Giorgis, *Jim Trelease's Read Aloud Handbook.*

8 Carl Erik Landhuis, Richie Poulton, David Welch, and Robert John Han-
   cox, "Does Childhood Television Viewing Lead to Attention Problems
   in Adolescence? Results from a Prospective Longitudinal Study," *Pediatrics* 120, no. 3 (2007): 535, https://pediatrics.aappublications.org/content
   /120/3/532?maxtoshow=&hits=10&RESULTFORMAT=&fulltext
   =TV%20Linked%20to%20Kids%20Attention%20Problems&searchid
   =1&FIRSTINDEX=0&sortspec=relevance&resourcetype=HWCIT.

9 Caroline J. Blakemore and Barbara Weston, *Every Word Counts* (Cre-
   ateSpace Independent Publishing Platform, 2016), 13.

10 Carol Copple and Sue Bredekamp, eds., *Developmentally Appropriate
   Practice* (Washington, DC: National Association for the Education of
   Young Children, 2010).

11 Diamond and Hopson, *Magic Trees of the Mind,* 88.

12 Blakemore and Weston, *Every Word.*

13 Susan Irvine, "Tips for Building a Child's Imagination," First Five
   Years, October 10, 2019, https://www.firstfiveyears.org.au/early-learning
   /tips-for-building-a-childs-imagination.

14 Martha Pennington and Robert Waxler, *Why Reading Books Still Matters: The Power of Literature in Digital Times* (Oxfordshire, England:
   Routledge, 2017).

15 "Reading Enhances Imagination," World Literacy Foundation, April 8, 2021,
   https://worldliteracyfoundation.org/reading-enhances-imagination/.

16 Kendra Cherry, "What Is Object Permanence?," Very Well Mind, April
   24, 2021, https://www.verywellmind.com/what-is-object-permanence
   -2795405.

2. Benefits of Reading Aloud for Language Acquisition

1   "Speech and Language Developmental Milestones," National Institute on Deafness and Other Communication Disorders, last modified March 6, 2017, https://www.nidcd.nih.gov/health/speech-and-language#2.

2   Mayo Clinic, "Language Development: Speech Milestones for Babies," February 23, 2022, https://www.mayoclinic.org/healthy-lifestyle /infant-and-toddler-health/in-depth/language-development/art-20045163.

3   Kathy Hirsh-Pasck, Jennifer M. Zosh, Roberta Michnick Golinkoff, James H. Gray, Michael B. Robb, and Jordy Kaufman, "Putting Education in 'Educational' Apps: Lessons from the Science of Learning," *Psychological Science in the Public Interest* 16, no. 1 (2015): 25.

4   Rachel R. Romeo, Julia A. Leonard, Sydney T. Robinson, Martin R. West, Allyson P. Mackey, Meredith L. Rowe, and John D. E. Gabriel, "Beyond the 30-Million-Word Gap: Children's Conversational Exposure Is Associated with Language-Related Brain Function," *Psychological Science* 29 (2018): 704.

5   Jessica Sidler Folsom, "Dialogic Reading: Having a Conversation About Books," Reading Rockets, 2017, https://www.readingrockets.org/article /dialogic-reading-having-conversation-about-books.

6   Folsom, "Dialogic Reading."

7   Trelease and Giorgis, *Jim Trelease's Read Aloud Handbook*, 25.

8   Betty Hart and Todd Risley, *Meaningful Differences in the Everyday Experience of Young American Children* (Baltimore: Brookes Publishing, 1996), 160, 198, 199.

9   Linnea C. Ehri, "Development of Sight Word Reading: Phases and Findings," in *The Science of Reading: A Handbook*, ed. Margaret J. Snowling and Charles Hulme (Hoboken, NJ: Blackwell Publishing, 2005), 135–154, https://doi.org/10.1002/9780470757642.ch8.

10 "Rare Words Are Easier to Learn than Common Words," ASHA Wire, February 1, 2013, https://doi.org/10.1044/leader.FTJ3.18022013.33.

11 Lauren L. Emerson, Nicole Loncar, Carolyn Mazzei, Isaac Treves, and Adele E. Goldberg, "The Blowfish Effect: Children and Adults Use Atypical Exemplars to Infer More Narrow Categories During Word Learning," *Journal of Child Language* 1 (2019), https://doi.org/10.1017 /S0305000919000266.

12 Vicky Bowman, "Eight Creative Ideas to Help Your Child Learn New Words," NAEYC, accessed April 13, 2022, https://www.naeyc.org /our-work/families/literacy/learn-new-words.

13 "Sharing Rich Read-Aloud Experiences," Scholastic, http://mediaroom .scholastic.com/files/readaloud-sharing.pdf.

3. Benefits of Reading Aloud for Social-Emotional Health

1 "Tips for Promoting Social-Emotional Development," Zero to Three official website, accessed April 11, 2022, https://www.zerotothree.org /resources/225-tips-for-promoting-social-emotional-development.

2 Nancy Eisenberg, Tracy L. Spinrad, and Natalie D. Eggum, "Emotion-Related Self-Regulation and Its Relation to Children's Maladjustment," *Annual Review of Clinical Psychology* 6 (2010): 495–525; Catherine E. Snow and Susan B. Van Hemel, eds., *Early Childhood Assessment: What, Why, and How* (Washington, DC: National Academies Press, 2008).

3 Tamara G. Halle and Kristen E. Darling-Churchill, "Review of Measures of Social and Emotional Development," *Journal of Applied Developmental Psychology* 45 (2016): 10, https://www.sciencedirect.com/science /article/pii/S0193397316300065.

4 Jill M. Raisor and Stacy D. Thompson, "Guidance Strategies to Prevent and Address Preschool Bullying," NAEYC, accessed April 13, 2022, https://www.naeyc.org/resources/pubs/books/excerpt-from-spotlight -social-emotional-development.

5 Diane E. McClellan and Lilian G. Katz, "Assessing Young Children's Social Competence," ERIC Digests, 2001, https://files.eric.ed.gov/fulltext /ED450953.pdf.

6 Gurdon, *Enchanted Hour.*

7 "Parentese Is the New Baby Talk That Will Help Your Baby Develop Speech," *USA Today*, February 4, 2020, https://www.usatoday .com/videos/life/parenting/2020/02/04/parentese-new-scientifically -approved-baby-talk-build-speech/4653619002/.

8 "Parentese Is the New Baby Talk," *USA Today.*

9 Scholastic, *Rise of the Read Aloud.*

10 Amy Joyce, "Why It's Important to Read Aloud with Your Kids and How to Make It Count," *Washington Post*, February 16, 2017, https://www .washingtonpost.com/news/parenting/wp/2017/02/16/why-its-important -to-read-aloud-with-your-kids-and-how-to-make-it-count/.

11 Afsaneh Moradian, "The Importance of Asking About Pronouns," Free Spirit, October 19, 2020, https://freespiritpublishingblog.com/2020/10/19 /the-importance-of-asking-about-pronouns/.

12 Natalia Kucirkova, "How Could Children's Storybooks Promote Empathy? A Conceptual Framework Based on Developmental Psychology and Literary Theory," *Frontiers in Psychology* 10 (2019): 121.

13 Susan Haynes, "Why It's Crucial to Talk to Kids About Gender Pronouns," *Time*, June 3, 2021, https://time.com/6053526/children -pronouns-gender-inclusivity/.

14 Scholastic, *Rise of the Read Aloud*.

15 Scholastic, *Kids & Family Reading Report*.

16 Scholastic, "Reading to Navigate the World," in *Kids & Family Reading Report*, https://www.scholastic.com/readingreport/navigate-the-world .html.

17 Scholastic, "Reading to Navigate."

18 Scholastic, "Reading to Navigate."

19 Scholastic Parents Staff, "A Love of Learning," Scholastic, accessed April 13, 2022, https://www.scholastic.com/parents/school-success/love -learning.html.

20 Scholastic, "Reading to Navigate."

Part II: The Six Stages of Reading Aloud: From Birth Through School Age

1 "How ASQ Works," ASQ, Paul H. Brookes Publishing Co., accessed April 13, 2022, https://agesandstages.com/about-asq/how-asq-works/.

2 "Examples of Developmental Stages," University of Washington, accessed April 13, 2022, https://depts.washington.edu/allcwe2/fosterparents/training /chidev/cd04.htm.

3 William Sears and Martha Sears, *The Baby Book* (New York: Little, Brown and Co., 1993); Diamond and Hopson, *Magic Trees of the Mind*; Roberta Michnick Golinkoff and Kathy Hirsh-Pasek, *How Babies Talk* (New York:

Penguin Books, 2000); William Staso, *Neural Foundations: What Stimulations Your Baby Needs to Become Smart* (Santa Maria, CA: Great Beginnings Press, 1995); William Staso, *Brain Under Construction: Experiences That Promote the Intellectual Capabilities of Young Toddlers* (Orcutt, CA: Great Beginnings Press, 1997).

4. The Listener: Pregnancy to 2 Months

1 Joyce, "Why It's Important," https://www.washingtonpost.com/news /parenting/wp/2017/02/16/why-its-important-to-read-aloud-with-your -kids-and-how-to-make-it-count/.

5. The Observer: 2 to 4 Months

1 "The Best Book Ever for Newborns!" (customer review of *Playtime, Maisy!*), Amazon.com, March 15, 2002, https://www.amazon.com /Playtime-Maisy-Lucy-Cousins/product-reviews/0763616028/.

10. Using Music to Support Read Alouds

1 Center for Effective Reading Instruction, "Phonological and Phonemic Awareness: Introduction," Reading Rockets, accessed April 13, 2022, https://www.readingrockets.org/teaching/reading101-course/modules /phonological-and-phonemic-awareness-introduction.

2 "Phonological Awareness," Massachusetts Department of Elementary and Secondary Education, last modified December 14, 2021, https:// www.doe.mass.edu/massliteracy/skilled-reading/fluent-word-reading /phonological-awareness.html.

3 "Phonological Awareness," University of Michigan, accessed April 13, 2022, http://dyslexiahelp.umich.edu/professionals/dyslexia-school /phonological-awareness.

12. Using Technology to Support Read Alouds

1 Jennifer F. Cross, "What Does Too Much Screen Time Do to Children's Brains?" Health Matters, accessed April 13, 2022, https://healthmatters .nyp.org/what-does-too-much-screen-time-do-to-childrens-brains/.

2 Caroline Gaudreau, Yemimah A. King, Rebecca A. Dore, Hannah Puttre, Deborah Nichols, Kathy Hirsh-Pasek, and Roberta Michnick Golinkoff,

"Preschoolers Benefit Equally from Video Chat, Pseudo-Contingent Video, and Live Book Reading: Implications for Storytime During the Coronavirus Pandemic and Beyond," *Frontiers in Psychology* 11 (2020): 2158.

3 Tiffany G. Munzer, Alison L. Miller, Heidi M. Weeks, Niko Kaciroti, and Jenny Radesky, "Differences in Parent-Toddler Interactions with Electronic Versus Print Books," *Pediatrics* 143 (2019).

4 "National Read Aloud Survey Shows Most Parents Are Not Reading Enough to Their Children," Durham's Partnership for Children, March, 11, 2016, https://dpfc.net/national-read-aloud-survey-shows-parents -not-reading-enough-children/.

5 Dan Seifert, "Amazon's Latest Alexa Trick Is Helping Kids Read," *Verge*, June 29, 2021, https://www.theverge.com/2021/6/29/22554428 /amazon-reading-sidekick-alexa-echo-skill-kids-voice-profiles.

6 Lois Bridges, "The Power of Reader's Choice and Identity," Bring Me a Book, accessed April 13, 2022, https://www.bringmeabook.org /the-power-of-readers-choice-and-identity/.

7 Joyce, "Why It's Important," https://www.washingtonpost.com/news /parenting/wp/2017/02/16/why-its-important-to-read-aloud-with-your -kids-and-how-to-make-it-count/.

Epilogue: Raising a Literacy-Rich Child

1 Lois Bridges, "The Power of Reader's Choice and Identity," Bring Me A Book, 2021, https://www.bringmeabook.org/wp-content/uploads /2021/06/BMAB_RESEARCH-ROUNDUP_Bridges_v4.pdf.

2 K. M. Edmunds and Kathryn L. Bauserman, "What Teachers Can Learn About Reading Motivation Through Conversations with Children," *Reading Teacher* 59, no. 5 (2006): 415; Nancy Atwell, *The Reading Zone* (New York: Scholastic, 2007).

3 Gurdon, *Enchanted Hour.*

INDEX